Wavelets and Fractals
in
Earth System Sciences

T0173049

Wavelets and Fractals
in
Earth System Sciences

Editors

E. Chandrasekhar
V. P. Dimri
V. M. Gadre

CRC Press
Taylor & Francis Group
Boca Raton London New York

CRC Press is an imprint of the
Taylor & Francis Group, an **informa** business
A CHAPMAN & HALL BOOK

MATLAB® is a trademark of The MathWorks, Inc. and is used with permission. The MathWorks does not warrant the accuracy of the text or exercises in this book. This book's use or discussion of MAT-LAB® software or related products does not constitute endorsement or sponsorship by The MathWorks of a particular pedagogical approach or particular use of the MATLAB® software.

CRC Press
Taylor & Francis Group
6000 Broken Sound Parkway NW, Suite 300
Boca Raton, FL 33487-2742

First issued in paperback 2019

© 2014 by Taylor & Francis Group, LLC
CRC Press is an imprint of Taylor & Francis Group, an Informa business

No claim to original U.S. Government works

ISBN-13: 978-1-4665-5359-0 (hbk)
ISBN-13: 978-0-367-37919-3 (pbk)

**Visit the Taylor & Francis Web site at
http://www.taylorandfrancis.com**

**and the CRC Press Web site at
http://www.crcpress.com**

Contents

Foreword vii

Preface...ix

Contributors.. xiii

1. **Introduction to Wavelets and Fractals**..1
   E. Chandrasekhar and V. P. Dimri

2. **Construction of Wavelets: Principles and Practices** 29
   Manish Sharma, Ashish V. Vanmali, and Vikram M. Gadre

3. **Genesis of Wavelet Transform Types and Applications** 93
   N. Sundararajan and N. Vasudha

4. **Multiscale Processing: A Boon for Self-Similar Data,**
   Data Compression, Singularities, and Noise Removal...................... 117
   Ratnesh S. Sengar, Venkateswararao Cherukuri, Arpit Agarwal, and
   Vikram M. Gadre

5. **Fractals and Wavelets in Applied Geophysics with**
   Some Examples.. 155
   R. P. Srivastava

6. **Role of Multifractal Studies in Earthquake Prediction**...................... 177
   S. S. Teotia and Dinesh Kumar

7. **Geomagnetic Jerks: A Study Using Complex Wavelets**...................... 195
   E. Chandrasekhar, Pothana Prasad, and V. G. Gurijala

8. **Application of Wavelet Transforms to Paleomonsoon Data**
   from Speleothems.. 219
   M. G. Yadava, Y. Bhattacharya, and R. Ramesh

9. **Unraveling Nonstationary Behavior in Rainfall Anomaly and**
   Tree-Ring Data: A Wavelet Perspective..229
   Prasanta K. Panigrahi, Yugarsi Ghosh, and Deepayan Bhadra

10. **Phase Field Modeling of the Evolution of Solid–Solid and**
    Solid–Liquid Boundaries: Fourier and Wavelet Implementations.... 247
    M. P. Gururajan, Mira Mitra, S. B. Amol, and E. Chandrasekhar

Index 273

Foreword

Wavelets were invented by the French geophysicist Jean Morlet, so it is most appropriate that the Department of Earth Sciences at the Indian Institute of Technology Bombay (IITB), with support from the Ministry of Earth Sciences and the Department of Space, organized an interesting meeting on wavelets and fractals in earth sciences at IITB early last year. The story of Morlet's invention is well known, and has been narrated in tributes given by Yves Meyer and Pierre Goupillaud. The wavelet idea had some precursors in the form of techniques to provide information on frequency scales in running time, as the human ear does so naturally in music and speech for its owner. An example of the kind of problem that such techniques attempt to handle is the decomposition of a signal whose frequency is constant over relatively short times but otherwise keeps varying throughout (as it happens familiarly in Indian music). Classical Fourier series cannot handle such inherently nonstationary phenomena. Morlet's achievement consisted of devising a system that was most appropriate for this class of problems; a 2-D transform of a 1-D transient, nonstationary signal could be defined, and furthermore, the relation could be inverted to reconstruct the original signal from the transform. The signals that Morlet was handling were those encountered in prospecting for oil, which was the business run by his employer; here, the waves transmitted into and reflected back from different layers in the soil return to the surface, with different frequencies at different times. Morlet's first article on the subject seemed so far-out and abstract to the geophysical journal he submitted it to that it was rejected. However, a collaboration with a theoretical physicist, Alex Grossmann, led to an elegant formulation that was mathematically convincing and sharply crystallized the wavelet idea.

Since then, work on wavelets has grown very rapidly. As often happens with new ideas, the first reactions of skepticism to Morlet's work were quickly followed by great appreciation after the publication of the Morlet–Grossmann article in 1984. There is now a whole series of wavelets of different kinds for different applications that range from data compression to image processing, and the detection of hidden order in apparently chaotic signals. Many books are now available on the subject—from relatively abstract mathematical treatments to books such as *Wavelets for Dummies*. The technique has been used not only in seismology and oil prospecting but also in meteorology, fluid dynamics, quantum physics, data analysis and compression, characterization of archeological structures or volcanic activity, and so on, to multiscale filtering and processing in electrical engineering. There are discrete and continuous transforms, 1-D, 2-D, and 3-D transforms, those that are good at detecting sharp lines and others in smoothing them out. The list is long.

This volume covers a great deal of this ground. Although it has a legitimate emphasis on geophysical applications, there are also chapters by electrical, aerospace, and materials engineers. The revolution that wavelets have wrought is that they provide, for the first time, a set of versatile tools that would be the default choice if today one is investigating transient non-stationary processes or data of any kind. This book should help to bring out the richness of wavelet (and fractal) methods now available, and introduce them to new entrants to research by offering a good mix of theory and specific applications in many different fields.

I wish to compliment the Indian geophysical and engineering communities for getting together to organize this very useful, timely and stimulating meeting.

Roddam Narasimha
Jawaharlal Nehru Centre for
Advanced Scientific Research
Bangalore

Preface

The subject of wavelet analysis and fractal analysis is fast developing and has drawn a great deal of attention in varied disciplines of science and engineering. Over the past couple of decades, wavelets, multiresolution analysis, and multifractal analysis have been formalized into a thorough mathematical framework and have found a variety of applications with a significant effect on several branches of earth system sciences, such as seismology, well-logging, potential field studies, geomagnetism and space magnetism, atmospheric turbulence, space–time rainfall, ocean wind waves, fluid dynamics, seafloor bathymetry, oil and gas exploration, and climate change studies among others. It is certain that there will be plenty of applications of wavelets and fractals in earth sciences in the years to come.

This book primarily addresses the important question "Why *not* another book?" instead of "Why another book" on wavelets and fractals? Commensurate with the rapid progress in the application of wavelets and fractals in all fields of science and engineering in general, and in earth sciences, in particular, the available books and other forms of literature on these novel signal analysis tools are alarmingly few. This has been the motivation for us to edit this book. Through this book, we attempt to highlight the role of such advanced data processing techniques in present-day research in various fields of earth system sciences. The book is composed of a unique collection of a wide range of application-oriented research topics in a multitude of fields in earth system sciences and presents the same under one umbrella.

The book consists of 10 chapters, providing a well-balanced blend of information about the role of wavelets, fractals, and multifractal analyses with the latest examples of their application in various research topics. We admit that the reader may find some basic concepts of continuous and discrete wavelet transformation techniques being repeated in some chapters. Given the different perspectives and wide applications of these techniques in different fields of science, it was felt necessary to retain this repetition for the sake of continuity in the text, particularly when such information is important and relevant to the research content of the chapter, for its better understanding. The chapters are written by experts in their respective fields. Starting from the fundamental concepts of the theory of wavelets and fractals, their application in a variety of fields in earth sciences is described. The construction of wavelets, second-generation wavelets, the role of fractals in earthquake prediction, the application of wavelets in geomagnetism, potential field studies, atmospheric rainfall anomalies, paleoclimate studies, and phase field modeling constitute the book.

In Chapter 1, the fundamental concepts of wavelets and fractals and their use in various branches of geosciences research are discussed albeit without many of the mathematical details. The concept of multifractals and their necessity for the effective understanding of complex geophysical phenomena such as earthquakes is also explained in this chapter. Finally, the relation between wavelets and fractals and the advantage of wavelet-based fractal analysis in studying the signals is discussed.

Chapter 2 builds up the whole philosophy of filter banks as a template for the construction of wavelets. Beginning with the Haar wavelet and using that as a platform to explain the close relation between filter banks and wavelets, the authors explain the need for "higher order" filter banks in wavelet analysis. Besides explaining the construction of some classes of orthogonal and biorthogonal wavelets, the authors present some of their own approaches for designing filter banks keeping the uncertainty or time–frequency bandwidth product as a criterion for optimization. Some designs are presented along with their performance in a compression application, which is very relevant to solving problems in the field of earth sciences.

Chapter 3 presents the genesis of wavelet transform. It discusses the development of second-generation wavelets and their construction. A comprehensive review of some applications of wavelet transform in the field of geophysics, particularly in potential fields, is highlighted.

Chapter 4 discusses multiresolution decomposition as a way to create multiscale relationships that help to identify the self-similarity and singularity properties in the data. It emphasizes the point: Why is multiscale representation very useful in studies such as data compression and noise removal? The role of wavelets in data compression is especially useful to store large volumes of geological maps, subsurface geophysical models, and remote sensing images.

Chapter 5 discusses some examples of the application of fractals and wavelets in geophysics. The most generally used "box-counting" method to determine the fractal dimension is detailed with an illustrative example of field data. An important application of wavelets to reduce the runtime for solving large matrices in gradient-based inversion of geophysical data is explained, as well as providing an example of the application of wavelets in analyzing potential field data.

Chapter 6 highlights the application of multifractals in earthquake prediction studies. Temporal variations of heterogeneity in seismicity using multifractal analysis have been reported in various tectonic regions for both microseismicity as well as macroseismicity data. By examining the generalized fractal dimension Dq and the spatiotemporal variation of Dq spectra, representative of seismicity pattern and their distribution in any region, the role of multifractals in earthquake prediction is highlighted.

Chapter 7 explains the role of wavelets in the field of geomagnetism. It introduces complex wavelets and their application in understanding the phase characteristics of geomagnetic jerks, which occur due to the differential fluid flow on the surface of the outer core. This study, carried out using global magnetic observatory data from approximately eight decades, helps in understanding their space–time and time–frequency localization characteristics. Geomagnetic jerks, having a period of about one year, manifest sudden changes in the slope of geomagnetic secular variations. Phase characteristics of geomagnetic jerks aid in better understanding the direction of outer core fluid motions responsible for the generation of geomagnetic jerks. This chapter also highlights the hemispherical differences observed in the nature of occurrences of geomagnetic jerks.

A unique application of wavelets in studying the paleomonsoon data from speleothems is detailed in Chapter 8. Stable oxygen isotope variations in cave calcite ($CaCO_3$, stalactites and stalagmites), collectively known as speleothems, are used for paleomonsoon reconstruction. In this chapter, the authors present wavelet transformation together with multitaper analysis of the same data, and explain how the use of different filters can help identify hidden signals pertaining to solar modulation of the monsoon.

Another unique application of wavelets is in studying tree-ring data that is helpful for understanding rainfall anomalies and climate studies, which is detailed in Chapter 9. In this chapter, the authors highlight the different periodicities that they have obtained in their analysis of rainfall data, and tree-ring data and the good correlation among them. They also discuss insights on their effect on various present-day global phenomena.

Application of wavelet transformation in the studies of the kind presented in Chapters 8 and 9 are very rare and thus this book enjoys the privilege of having such rare discussions on such important areas of research, which are beneficial for monsoon and climate change studies.

Finally in Chapter 10, another unique application of wavelet transformation, namely, that of solving nonlinear partial differential equations (specifically, those from a phase field model), is presented. Phase field models are widely used in the field of materials science and engineering to study microstructures and to understand the solid–solid and solid–liquid boundaries. By carrying out a suitable nondimensionalization, these models can also be used to study boundaries at macro scales such as crust–mantle (solid–solid) and mantle–core (solid–liquid) boundaries. In this chapter, the authors describe how phase field models based on the Cahn–Hilliard (CH) equation can be used to understand the phase transition dynamics and the resultant (stress assisted) microstructural evolution of the interface between a non-hydrostatically stressed mineral and its solution. They also present results from a Fourier spectral implementation for the two-dimensional CH equation (with elastic stress effects), and a wavelet-collocation implementation for the one-dimensional CH equation (without elastic stress effects).

Some key features of the book are

- Prior knowledge of wavelets and fractals is not required to understand this book because each chapter introduces the needed concepts of wavelets and fractals.

- Focused set of applications centered on themes of thrust areas of present-day research in science and engineering.
- Healthy combination of basics and advanced material.
- Serves as excellent introductory material and also as an advanced reference text for students and researchers.

We place on record our sincere thanks to the Ministry of Earth Sciences, Government of India, ISRO (GBP) and ISRO (RESPOND) of the Department of Space, Government of India, for financial support. We express our sincere thanks and deep sense of gratitude to Prof. Roddam Narasimha, FRS, for readily agreeing to write the Foreword for this book and for his timely advice and moral support. We also express our sincere thanks to all the contributors of the chapters and the reviewers for helping us with their constructive criticism and meticulous reviews of the chapters. Finally, we thank the staff of CRC Press, Taylor & Francis group, particularly, Aastha Sharma, Marsha Pronin, and Rachel Holt and others for their kind cooperation and support throughout the evolution of this book.

E. Chandrasekhar
V. P. Dimri
Vikram M. Gadre

MATLAB® is a registered trademark of The MathWorks, Inc. For product information, please contact:

The MathWorks, Inc.
3 Apple Hill Drive
Natick, MA 01760-2098 USA
Tel: (508) 647-7000
Fax: (508) 647-7001
E-mail: info@mathworks.com
Web: http://www.mathworks.com

Contributors

Arpit Agarwal
Department of Electrical
  Engineering
Indian Institute of Technology
  Bombay
Mumbai, India

S. B. Amol
Department of Metallurgical
  Engineering and Material Science
Indian Institute of Technology
  Bombay
Mumbai, India

Deepayan Bhadra
Department of Instrumentation and
  Control Engineering
National Institute of Technology
Tiruchirappalli, India

Y. Bhattacharya
Center for Solar-Terrestrial Research
New Jersey Institute of Technology
Newark, New Jersey

E. Chandrasekhar
Department of Earth Sciences
Indian Institute of Technology
  Bombay
Mumbai, India

Venkateswararao Cherukuri
Department of Electrical
  Engineering
Indian Institute of Technology
  Bombay
Mumbai, India

V. P. Dimri
CSIR-National Geophysical
  Research Institute
Uppal Hyderabad, India

Vikram M. Gadre
Department of Electrical
  Engineering
Indian Institute of Technology
  Bombay
Mumbai, India

Yugarsi Ghosh
Department of Electronics and
  Communication Engineering
National Institute of Technology
Tiruchirappalli, India

V. G. Gurijala
Department of Earth Sciences
Indian Institute of Technology
  Bombay
Mumbai, India

M. P. Gururajan
Department of Metallurgical
  Engineering and Material Science
Indian Institute of Technology
  Bombay
Mumbai, India

Dinesh Kumar
Department of Geophysics
University of Kurukshetra
Kurukshetra, India

Mira Mitra
Department of Aerospace
    Engineering
Indian Institute of Technology
    Bombay
Mumbai, India

Prasanta K. Panigrahi
Department of Physics
Indian Institute of Science
    Engineering Research (IISER)
    Kolkata
Kolkata, India

Pothana Prasad
Department of Earth Sciences
Indian Institute of Technology
    Bombay
Mumbai, India

R. Ramesh
Geosciences Division
Physical Research Laboratory (PRL)
Ahmedabad, India

Ratnesh S. Sengar
Bhabha Atomic Research Centre
Mumbai, India

Manish Sharma
Department of Electrical
    Engineering
Indian Institute of Technology
    Bombay
Mumbai, India

R. P. Srivastava
National Geophysical Research
    Institute
Uppal, Hyderabad, India

N. Sundararajan
Department of Earth Science
Sultan Qaboos University
Muscat Sultanate of Oman

S. S. Teotia
Department of Geophysics
University of Kurukshetra
Kurukshetra, India

Ashish V. Vanmali
Department of Electrical
    Engineering
Indian Institute of Technology
    Bombay
Mumbai, India

N. Vasudha
Department of Mathematics
Vasavi College of Engineering
Hyderabad, India

M. G. Yadava
Geosciences Division
Physical Research Laboratory (PRL)
Ahmedabad, India

1

Introduction to Wavelets and Fractals

E. Chandrasekhar and V. P. Dimri

CONTENTS

Introduction 1
  Wavelets 1
  Fractals 4
Basic Theory and Mathematical Concepts of Wavelets and Fractals 5
  Wavelets 5
    Some Properties of Wavelets 7
    Continuous Wavelet Transformation 8
    Discrete Wavelet Transformation 9
  Fractals in Time Series Analysis 11
    Different Methods to Estimate Fractal Dimensions 13
Geophysical Significance of Wavelets and Fractals 16
  Wavelets in Geophysics 16
  Fractals in Geophysics 18
Relation between Wavelets and Fractals 18
  Methodology for Wavelet-Based Fractal Analysis 20
Conclusions 20
Acknowledgments 21
References 21

Introduction

Wavelets

The early 1800s saw a revolution in the understanding of mathematical functions when Jean Baptiste Joseph Fourier, in his groundbreaking observations, proclaimed that any mathematical function can be represented as a combination of several sines and cosines. This way, because the underlying frequencies of sine and cosines were known, it became easier to distinguish the frequencies of interest present in the signal under investigation. With the advent of subsequent thorough mathematical formalism for this transformation, the Fourier transform (FT) came into existence and is still used extensively to solve a variety of problems in science and engineering.

The sine and cosine functions are the basis functions in Fourier theory, which for any frequency is infinitely oscillatory in time and thus is *not* compactly supported (i.e., mathematically, the Fourier basis functions do not belong to the $L^2(R^n)$ functional space). As a result, Fourier theory can determine only the presence of all frequencies in the signal and cannot estimate when in time these frequencies of interest occur (also known as time localization). For example, if you have three signals in time-domain, with the first signal having a low-frequency part followed by a high-frequency part (Figure 1.1a), the second signal having a high-frequency part followed by a low-frequency part (Figure 1.1b), and the third signal is a combination of both these frequencies (Figure 1.1c), then the Fourier transformation in all three cases will yield the same spectra (barring the small ripples that you observe in the first two cases in their FT spectra that arise because of Gibbs phenomenon due to sudden frequency transitions; Figure 1.1). Now, the question is, how can the spectra of three different signals be the same? The answer is, this is the limitation of FT. This explains that FT can only tell us what frequencies are present in the signal and their respective average amplitude. Nothing more, nothing less. In other words, with FT, only the frequency localization is possible, but not the time localization.

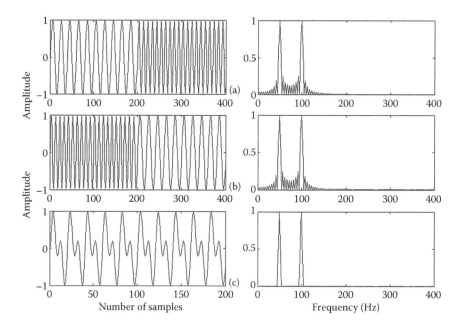

FIGURE 1.1
Example of three different signals with similar Fourier spectra. The left panels show (a) low-frequency signal appended with its high-frequency counterpart, (b) high-frequency signal appended with its low-frequency counterpart, and (c) a combination of both frequencies shown in (a) and (b). Their respective Fourier spectra are shown in the right panels.

However, it is becoming increasingly important and necessary to study some signals based not only on the frequency content but also on their time of occurrence (also called time–frequency localization) for a correct understanding and interpretation of the physical processes that generate such signals.

For example, knowledge of the behavior of the Earth's magnetic field variations at different times over different regions is always important to understand the spatiotemporal behavior of certain frequency characteristics of these variations (see Chapter 7 of this book and references therein). Similarly, in geophysical well-log data analysis, because different subsurface rock formations have different physical properties, which in turn reflect respective frequency characteristics, it becomes important to identify the locations of these frequencies (read formations) as a function of depth (space–frequency localization), particularly if these formations are oil-bearing/gas-bearing zones (see for example, Chandrasekhar and Rao).

Recognizing the drawback of FT, Gabor (1946) developed its small variant, called windowed FT (WFT) or short-time FT. The underlying principle of this technique is to break up the signal into smaller sections with fixed window length and Fourier analyze each section for its frequency content. Mathematical representation of the WFT of a time function $f(t)$ and a window function $g(t)$ is given by

$$\mathrm{WFT}_g(\tau, s) = F_g(\tau, s) = \int_{-\infty}^{\infty} f(t)g(t - \tau)e^{-i2\pi st}\, dt \tag{1.1}$$

$F_g(\tau, s)$, which is essentially the FT of $f(t)$ and $g(t - \tau)$, represents the amplitude and phase of the signal over time and frequency. $|F_g(\tau, s)|^2$ gives the spectrogram of $f(t)$. In WFT, although the short lengths of data are considered at each step, the main transformation again is FT, whose basis functions, as mentioned above, are infinitely supported and therefore the problem of time–frequency localization still persists. Furthermore, because the window length is fixed at each step in WFT, the resolution in the entire time–frequency plane will be the same (see Graps 1995; Mallat 1999). This is the drawback of WFT. Alternatively, although a wider window gives good frequency resolution, a narrower window gives good time resolution (Polikar 2001), indicating that they are governed by the uncertainty relation. Hence, WFT also was found not to be suitable where time–frequency or space–frequency localization is important.

Later, Banks (1975), with his complex demodulation technique, came out with some improvements in time–frequency localization of long-period geomagnetic data. This technique helped to determine the instantaneous power levels of the required frequency in the selected frequency band as a function of time. These are known as demodulates. Because each demodulate represents the time-local estimate of amplitude and phase of selected frequency in the entire data sequence, they can be treated as independent entities and thus it becomes very useful to examine each demodulate separately for their time–frequency characteristics in a given time series (Chandrasekhar 2000). Despite providing the instantaneous power spectrum and phase, the main drawback of this technique is that it does not have constant Q frequency resolution and also lacks a thorough mathematical framework (Zhang and Paulson 1997). The Q factor of a window (also called the fidelity factor) defines the ratio of center frequency of the filter to its bandwidth.

The uncertainty relation explains that, to have a better resolution of time and frequency, there must be a window function defined whose width should be allowed to vary inversely with the frequency. This facilitates in having a constant Q frequency resolution, and thus the wavelet is born! Oscillatory short waves, having finite amplitude and finite time duration, are known as *wavelets*. These time-limited waves can be successfully utilized to analyze the signals under investigation and provide not only the effective time–frequency localization of signals but also identify the hidden discontinuities or spikes of interest in the given signals. Wavelets thus facilitate the correct interpretation and understanding of the spatiotemporal behavior of various signals, which otherwise, is not possible with the FT.

Fractals

Fractal theory was not invented in a day by Benoit B. Mandelbrot, who coined this term to open a new discipline of mathematics that deals with nondifferentiable curves and geometrical shapes. Until Karl Weierstrass presented his work at the Royal Prussian Academy of Sciences on July 18, 1872, mathematics was confined to the study of differentiable functions. Weierstrass showed that for a positive integer a and $0 < b < 1$, the analytical function

$$\sum_{n=1}^{\infty} b^n \cos(a^n \pi x) \tag{1.2}$$

was not differentiable.

There are a few others, whose works are closely related to fractal theory, such as Von Koch, Julia, Fatou, and Hausdorff, who laid the foundations that were generously exploited by Mandelbrot to establish a fascinating branch of mathematics to analyze the natural geometrical shapes and more precisely quantify the length, area, and volume of these shapes. It was a landmark essay entitled, "How long is the coast of Britain? Statistical self-similarity and fractional dimension," written by Mandelbrot (1967), in which he related the work of his predecessors to real-world problems, namely, coastlines, clouds, mountains, and many other natural shapes, which he claimed are self-similar* and devised a method to quantify the geometrical properties of these shapes with the help of a dimension, which he termed as the fractal dimension. It is important to mention that as stated above, the fractal theory provides a framework to study irregular shapes as they exist, instead of approximating them using regular geometry, as has been done hitherto. For example, earth is approximated as a sphere, coastlines are approximated with straight lines, and many other natural shapes to the corresponding nearby (or closer) regular geometry, simply for the sake of mathematical convenience.

The concept of fractals is not confined to the study of geometrical shapes, as it might seem from the above introduction, rather it has been extended to the analysis of numerical data, for example, the variation of some physical properties of the earth with depth, spatial, and time series data from various branches of science including complex and extreme events (Sharma et al. 2012).

In the following sections, we provide a brief description of the mathematical concepts of wavelets and fractals, and their superiority over the existing signal processing tools, particularly, FT. We briefly discuss the methodology of continuous and discrete wavelet transformation and fractal analysis, whereas more details of these techniques are given in other chapters. Finally, we discuss the geophysical significance of these novel data analysis tools.

Basic Theory and Mathematical Concepts of Wavelets and Fractals

Wavelets

The word *wavelet* refers to a small wave, having finite length in space (ideally, functions representing such waves will be zero outside a finite time interval; also known as *compactly supported*) and is represented as a real-valued function, $\psi(t)$, adhering to the following conditions (Percival and Walden 2000)

$$\int_{-\infty}^{\infty} \psi(t)dt = 0 \qquad (1.3)$$

and

---

* If an object retains its original shape even after decomposing into smaller parts, then that object is said to have self-similar characteristics and is a fractal. Examples: Cantor set, Julia set, Cauliflower, cobweb, etc.

$$\int_{-\infty}^{\infty} \psi^2(t)\,dt = 1 \tag{1.4}$$

Equation 1.3 defines the *admissibility condition*, which explains that the wavelet function, $\psi(t)$, must be oscillatory in time and that it integrates to zero (i.e., having zero average). Equation 1.4 defines the *regularity condition*, which explains that the wavelet function, $\psi(t)$, must be finite in length and is square integrable (having finite energy). The wavelet function, $\psi(t)$, signifying the time–frequency localization is given by

$$\psi_{\tau,s}(t) = \frac{1}{\sqrt{s}}\,\psi\!\left(\frac{t-\tau}{s}\right) \tag{1.5}$$

where s (>0) indicates the scale and τ indicates the translation parameter. Equation 1.5 facilitates the provision to allow the width of the window function s to vary inversely with the frequency. Accordingly, s is analogous to frequency, in the sense that higher scales correspond to the low frequency content of the signal and lower scales correspond to high frequency content of the signal. The frequency–scale relation is given by $f = \dfrac{f_c}{s\Delta t}$, where f_c denotes the central frequency of the wavelet, Δt the sampling interval, and f the frequency corresponding to the scale. The translation parameter τ is linked to the time location of the wavelet function. As the wavelet is dilated and shifted during its operation on the signal (see subsections on Some Properties of Wavelets), it provides time-scale information in the transformed domain. A number of wavelets satisfy the conditions prescribed in Equations 1.3 and 1.4, and thus they form what is called "wavelet families." The function $\psi(t)$ is called "analyzing wavelet" or "mother wavelet." More details about the fundamentals of wavelet theory can be found in Daubechies (1992) and Mallat (1999, and references therein).

The continuous wavelet transformation (CWT) of a function, $f(t)$, is the result of its inner product with the wavelet function (Equation 1.3), given by

$$\mathrm{CWT}_{\tau,s} = \frac{1}{\sqrt{s}}\int f(t)\cdot\psi\!\left(\frac{t-\tau}{s}\right)dt \tag{1.6}$$

where, $f(t)$, $\psi(t) \in L^2(R)$. Equation 1.6 explains that the wavelet transformation gives a measure of the similarity between the signal and the wavelet function. Such a measure at any particular scale s_0 and translation τ_0, is identified by a wavelet coefficient. The larger the value of this coefficient, the higher the similarity between the signal and the wavelet at (τ_0, s_0) and vice versa. Higher wavelet coefficients indicate the high degree of suitability of the wavelet to study the signal. Another important point to note here is that, when s is close to zero, the CWT coefficients at that scale characterize the location of the singularity (if present) in the signal, in the neighborhood of τ. Central to this is its application to detect sudden jumps (transients) and/or discontinuities in the data and analyze fractals (see Mallat 1999).

From Equation 1.6, it can be understood that the CWT of a one-dimensional function is two-dimensional. Furthermore, it can also be shown that the CWT of a two-dimensional function is four-dimensional, and so on.

Some Properties of Wavelets

Vanishing Moments

The regularity condition (Equation 1.4) imposes an additional constraint on the wavelet function and makes the wavelet transform decrease quickly with decreasing s. This can be better explained with the concept of vanishing moments of the wavelet. If the wavelet coefficients for mth order polynomial are zero, then that wavelet is said to have m vanishing moments. This means that any polynomial up to order $m - 1$ can be represented entirely in scaling function space. In other words, a wavelet has m vanishing moments if its scaling function (also known as "father wavelet") can generate polynomials to a degree $\leq m - 1$ (see section on Discrete Wavelet Transformation for more details about scaling function). Alternatively, another simple way of defining wavelet vanishing moments is if the FT of the wavelet $\psi(t)$ is m times continuously differentiable, then the wavelet is said to have m vanishing moments (see Chandrasekhar and Rao 2012). Together with regularity condition, vanishing moments of the wavelet decide the fast decay of the wavelet.

Compact Support

This property explains that the wavelet vanishes outside a finite time interval. Shorter intervals indicate higher compactness of the wavelet and vice versa.

Translational Invariance

Another important property of wavelets is the translational invariance property, which explains that even a small time shift in the wavelet function results in a corresponding shift in the CWT output. Let us examine this.

Let $f_{\delta t}(t) = f(t - \delta t)$ be the translation of $f(t)$ by a small time shift, δt. The CWT of $f_{\delta t}(t)$ is

$$\mathrm{CWT}f_{\delta t}(\tau, s) = \frac{1}{\sqrt{s}} \int_{-\infty}^{\infty} f(t - \delta t)\psi\left(\frac{t - \tau}{s}\right) dt$$

$$= \frac{1}{\sqrt{s}} \int_{-\infty}^{\infty} f(t')\psi\left(\frac{t' - (\tau - \delta t)}{s}\right) dt \text{ where } t' = t - \delta t \qquad (1.7)$$

$$= \mathrm{CWT}f(\tau - \delta t, s)$$

Since the output is shifted the same way as the input signal, the CWT is translation-invariant. Similarly, the shift-invariant property for discrete wavelet transssform (DWT), in which the shifts are in dyadic scales can also be shown. For further details, the reader is referred to Ma and Tang (2001).

The discussion that we have had thus far on wavelets directly relates to the information that you will find in the following chapters. Hence, we limit our discussion to these aspects only. More details about different types of wavelets, their properties and applications, can be found in the respective chapters that follow. The reader is also advised to look into some fundamental books by Mallat (1999), Daubechies (1992), Chui (1992), Kaiser (1994), and Percival and Walden (2000), to cite a few.

Continuous Wavelet Transformation

In CWT (Equation 1.6), the signal to be transformed is convolved with the mother wavelet and the transformation is computed for different segments of the data by varying τ and s. Because the wavelet window can be scaled (shrunk or expanded) at different levels of analysis, the time localization of high-frequency components of the signal and frequency localization of low-frequency components of the signal can be effectively identified. In wavelet analysis, the wavelet window is first placed at the start of the signal. The signal and wavelet are convolved and the CWT coefficients are estimated. Next, the wavelet is translated (shifted in time) along the signal by a small amount and the CWT coefficients are again computed. This process is repeated until the end of the signal is reached. Next, the wavelet window is dilated (or stretched) by a small amount and placed at the start of the signal and the above process is repeated at that scale, at all translations. Likewise, wavelet spectra are calculated for many scales. Figure 1.2 describes the pictorial representation of the above operation. Finally, all the CWT coefficients computed at different dilations and translations are expressed in the form of scalograms, explaining the time-scale representations of the signal. Because of this, the wavelets are known as "mathematical microscopes."

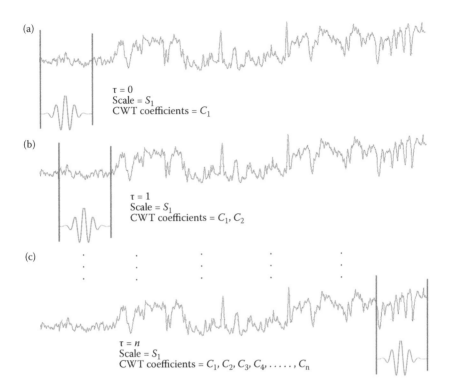

FIGURE 1.2
Pictorial representation of stepwise calculation of CWT coefficients. First, the CWT coefficients, when the wavelet with a translation $\tau = 0$ and dilation $s = 1$ are computed (a). Next, the coefficients, when the wavelet with a translation $\tau = 1$, without changing the dilation are obtained (b). Likewise, for the same "s" of the wavelet, the coefficients, $C_1 \ldots C_n$ are obtained. Next, the steps described in (a) through (c) are repeated for different dilations of the wavelet. Then, a complete scalogram depicting the time-scale representation of the signal under study is obtained.

Discrete Wavelet Transformation

Although the transformation obtained by shifting the wavelet continuously along the signal is called CWT, the transformation obtained by shifting the wavelet in discrete steps, with step-length equal to 2^j ($j = 1, 2, \ldots n$), is called DWT. It is important to note here that DWT will also be translational invariant (see subsection on Translational Invariance), only if the translations are in steps of 2^j ($j = 1, 2, \ldots n$), in which case it is called dyadic wavelet transformation. In DWT, a discrete set of wavelet scales and translations are used. In fact, the DWT operation is similar to digital filtering. DWT is computed using Mallat's algorithm (Mallat 1989), which essentially employs successive low-pass and high-pass filtering. First, the signal ($X[n]$) to be transformed is decomposed into high-frequency ($H[n]$) and low-frequency components

(*L*[*n*]) (also known as detailed and approximate coefficients, respectively). The detailed coefficients are known as level 1 coefficients. Then the approximate coefficients (low-frequency components) are again decomposed into the next level of detailed and approximate coefficients, and so on. Figure 1.3 depicts the DWT tree; a pictorial representation of the DWT process. At this point, it is important to understand exactly what happens at each level of decomposition when the signal under investigation is divided into low-frequency and high-frequency components. First, this enhances the resolution in frequency, thereby reducing its uncertainty by half. Accordingly, each of the filtered bands will have a band occupancy of $\omega/2$, which can be sampled at ω (cf. Nyquist theorem and its band pass extension). This decimation by 2 halves the time resolution as the entire signal after the first level of decomposition is represented by only half the number of samples. Thus, in the second level of decomposition, while the half-band low-pass filtering removes half of the frequencies, the decimation by 2 doubles the scale. With this approach, while the time resolution in the signal gets improved at high

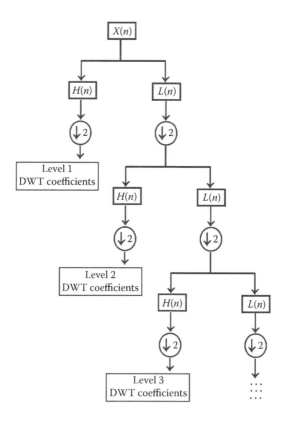

FIGURE 1.3
The DWT tree. The number "2" in the circles represents the decimation factor. See text for a detailed description of this tree.

frequencies, the frequency resolution is improved at low frequencies. Such a process is repeated until the desired levels of resolution in frequency are reached. However, the important point to note here is that at every wavelet operation, only half of the spectrum is covered. This means an infinite number of operations are needed to complete DWT. The solution to this problem does not lie in computing the DWT all the way down to zero with wavelet spectra, but to calculate up to a reasonable level, when it is small enough. The scaling function of the wavelet is used to check this. The scaling function is just a signal with a low-pass spectrum. The entire DWT operation can be summarized as follows: In DWT, if we analyze a signal using the combination of scaling function and wavelet function, the scaling function contributes to the calculation of the low-pass spectrum, whereas the rest is done by the wavelet function. The DWT of the original signal is then obtained by concatenating all the coefficients, $H[n]$ and $L[n]$, starting from the last level of decomposition. More details regarding DWT are given in Chapters 2, 3, and 4.

Fractals in Time Series Analysis

Often, geophysical data are in the form of time–space series (Dimri 1992). It has a combination of stochastic, trend, and periodic components (Malamud and Turcotte 1999). To know the stochastic component, one must compute the statistical distribution of values and persistence. The most commonly used method to determine persistence is spectral analysis, in which the power spectrum of a time series is plotted against the frequency (or wave number in case of a space series) and the value of a slope, known as the scaling exponent (e.g., β), giving an estimation of persistence (Dimri et al. 2012), which could be weak or strong. There exists a relation between β, the Euclidean dimension (E), and fractal dimension (D) of data as

$$D = E - 1 - \beta \qquad (1.8)$$

Persistence of the time series is a measure of correlation between adjacent values of the time series. The uncorrelated or random data with zero persistence is the white noise or $\beta = 0$. The time series is antipersistent if adjacent values are anticorrelated, that is, $\beta < 0$. The time series is persistent if adjacent values are positively correlated with each other. For $\beta > 0$, the series is positively correlated.

The density, susceptibility distribution, and reflectivity sequence of many boreholes around the globe have been investigated. The power spectral density of density, susceptibility distributions, and reflectivity sequences of some of the boreholes are shown to follow a fractal distribution (Dimri 2000a). The power spectrum of susceptibility distributions from 4 km depth of the German Continental Deep Drilling Project (KTB) was computed by

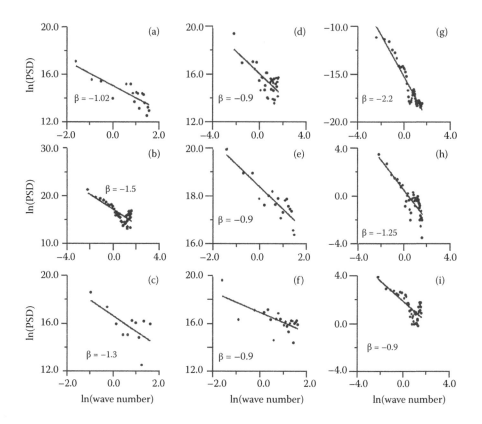

FIGURE 1.4
Log-log plot of wave number versus psd (in db) (a) for ODP Borehole complex impedance data; (b) of resistivity data for depth range one, KTB (VB); (c) of resistivity data for depth range two, KTB (VB); (d) of resistivity data for depth range three KTB (VB); (e) of resistivity data for depth range 1.6 to 2 km, of KTB (VB) a test case; (f) of resistivity data for depth range 2.0 to 2.5 km, of KTB (VB) a test case; (g) of conductivity data of KTB (HB); (h) of chargeability data for depth range one of KTB (VB); and (i) of chargeability data for depth range three of KTB (VB). (After Vedanti, N., and Dimri, V.P., *Indian J. Mar. Sci.* 32(4), 273–278, 2003.)

Maus and Dimri (1995a). It can also be observed in Figures 1.4 and 1.5 that there exists a fractal behavior in the electrical and thermal properties of the continental crust (Vedanti and Dimri 2003; Vedanti et al. 2011). From these investigations, it can be clearly seen that the power spectrum is not flat but rather it follows some scaling laws.

Now, once it is known that the physical properties follow a fractal behavior, it entails the reformulation of the mathematical formulae in frequency domain for existing techniques in geophysics that have been derived based on the Gaussian/random distribution of the physical properties. Such an approach was first developed by Spector and Grant (1970) and Naidu (1970), and was used to estimate the thickness of sedimentary basin from gravity and magnetic data. This method is known as the spectral analysis

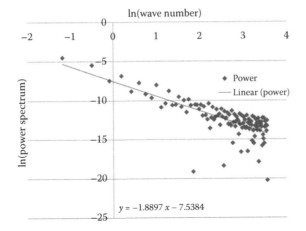

FIGURE 1.5

Log-log plot of power spectrum of heat production data versus wave number for the Hidaka meta-morphic belt (Hokkaido, Japan). Best linear fit gives a scaling exponent of 1.9, indicating the presence of fractal behavior in the data set. Nature of linear fit indicates strong correlation between the adjacent points and substantiate power–law regime in the data. (After Vedanti et al. 2001).

method and was also used by Negi et al. (1986) to estimate the thickness of the basalt in the Deccan Volcanic Province. However, for the fractal distribution of physical properties, a new method called scaling spectral method was developed and applied for various case studies of potential field data (Pilkington and Todoeschuck 1993; Maus and Dimri 1994, 1996; Fedi et al. 1997; Dimri 2000c; Bansal and Dimri 1999, 2005; and Bansal et al. 2006).

Different Methods to Estimate Fractal Dimensions

The essence of fractal analysis is the estimation of fractal dimension. With the use of fractal dimension, one can measure the length of a curve or area of rugged boundaries, or the volume of uneven surfaces, more precisely. Before the use of fractal dimensions, one was forced to use the Euclidean concept in which geometrical shapes consisting of straight-line polygons have the integer dimension values 1, 2, and 3 for a line, area, and volume, respectively. Mathematically, the length, L of a curve is a function of the size of the measuring instrument, for example, a scale of length ε. The relationship between L and ε is given by

$$L = A \, (1/\varepsilon)^d \tag{1.9}$$

where A is a constant and d is the "measured fractal dimension." Evaluating the logarithm of Equation 1.9 on both sides, the parameter, "d" can be obtained by

$$\log L = \log A + d \, \log(1/\varepsilon) \tag{1.10}$$

Therefore, if we plot log L versus log $(1/\varepsilon)$, we get a straight line whose slope gives the required value for d. One can show that the measured dimension, d, is related to the conventionally defined fractal dimension D by the relation:

$$D = 1 + d \tag{1.11}$$

If we calculate d for the coastline of Britain using the above procedure, we get a value of 0.28. Thus, we obtain the fractal dimension as 1.28. Therefore, the fractal dimension of the coastline of Britain is a fractional number (i.e., not an integer) between 1 and 2. The important point is that the fractal dimension is greater than the topological dimension (the topological dimension of a one-dimensional object is 1, in this case, the coastline). On the other hand, if we measure D for objects such as circles, we find that fractal dimension equals the topological dimension (both are equal to 1). Natural objects such as coastlines, exhibit fractal characteristics only over a limited range of scales, that is, the straight line behavior in the log-log plot used to calculate d holds only for a certain range of ε, that is, measuring length (Rangarajan 2000).

For complex cases such as heterogeneity in seismicity and fluid turbulence (Frisch and Parisi 1985), a single fractal dimension cannot explain the process. Such cases are best modeled using multifractals. Moharir (2000) demonstrated that the sum of two fractal processes is not a fractal. In such a case, a generalized fractal dimension, D_q, as a function of q, which varies from $-\infty$ to $+\infty$ was introduced. Hentshel and Procaccia (1983) presented D_q as

$$D_q = (1/q - 1)\lim_{r \to 0}\left[\log\left(\sum \{P_i(r)\}^q\right)/\log r \right] \tag{1.12}$$

where $P_i(r)$ is the probability that the events fall into a box with a size r. In the case of $q = 0$, 1, and 2: D_0 coincides with the capacity dimension (D_c); D_1 with the information dimension (D_i) and D_2, to the correlation dimension (D_{corr}). In general, the relation $D_0 > D_1 > D_2 > ... D_\infty$ occurs within D_q. The multifractal analysis of complex phenomena involves estimation of the generalized dimension D_q first and the $D_q - q$ relation curve. There are various methods to determine D_q as given below.

Box Counting Method

It is a very popular method to estimate fractal dimension. In this method, the data are initially superimposed on a square grid of size r_0. The unit r_0^2 is sequentially divided into small squares of size $r_i = r_0/2, r_0/4, r_0/8$, etc. The number of squares $N(r_i)$ intersected by at least one feature is counted each time. If the system is a self-similar structure, the relation between $N(r_i)$ and fractal dimension D is given by

$$N(r_i) \propto \left(\frac{r_0}{r_i} \right)^D \tag{1.13}$$

where D is interpreted as the fractal dimension of the lineament/drainage system. The fractal dimension D can be determined from the slope of the plot of $\log N(r_i)$ versus $\log (r_0/r_i)$.

Fixed-Size Algorithm

This is a generalization of the correlation method. In this method, the estimation of scaling of mass (e.g., number of earthquakes) within the circle of radius r increases with r. The dimension D_q is estimated from the scaling of mass with size for fixed-sized circles:

$$\log \langle M (<r)^{q=1} \rangle \approx (q-1) D_q \log r \tag{1.14}$$

Here, $M(<r)$ is the mass within the fixed radius, $\langle \rangle$ means the average of the mass for fixed size of the circle. The mass can be the number of data sets under consideration. The method is reported to be effective for determining the spectrum for $q > 0$ (Grassberger et al. 1988), but it is unstable for negative q when the data are limited.

Fixed-Mass Algorithm

This is also a generalization of the correlation method. The algorithm is the same as that used in the fixed radius method. Here, the smallest radius within which a fixed mass m can be included increases as the mass increases. The relationship to be used is

$$\log \langle R (< m)^{-(q-1) D} \rangle \approx -(q-1) \log m \tag{1.15}$$

The notations are similar to Equation 1.14. This method is reported to be superior to the previous one for calculating the spectrum for negative q (Grassberger et al. 1988). The applications for earthquake data are shown by several other workers (for example, see Hirata 1989; Hirabayashi et al. 1992).

Correlation Integral Method

In this method, the local density function $n_i(r)$ is defined by the following equation:

$$n_i(r) = \left(\frac{1}{N-1} \right) \left\{ \lim \sum \left(r - |x_i - x_j| \right) \right\} \tag{1.16}$$

where N is the number of data points recorded with a temporal resolution period, $(r - |x_i - x_j|)$ is the Heaviside function that counts how many pairs of points (x_i, x_j) fall within the scaling radius, r.

$$(r - |x_i - x_j|) = 0, \text{ if } (r - |x_i - x_j|) < 0 \tag{1.17}$$

$$= 1 \text{ if } (r - |x_i - x_j|) > 0 \tag{1.18}$$

To study the multifractal behavior of time series of earthquakes, r is replaced by scaling time t (Wang and Lee 1996) and ($|x_i - x_j|$) by the inter-occurrence time ($|t_i - t_j|$). Therefore, a generalized correlation integral $C_q(t)$ for the interoccurrence time t is defined by the equation

$$C_q(t) = \left\{ \left[\sum (n_i(t))^{q-1} \right] \Big/ N \right\}^{1/q-1} \tag{1.19}$$

Using this correlation integral, the temporal generalized fractal dimension D_q is then defined by the scaling relation as

$$Cq(t) \sim tDq \tag{1.20}$$

The generalized fractal dimension Dq is determined by plotting $Cq(t)$ versus t on a log-log graph. Applications of multifractals to earthquake data have been shown by Sunmonu et al. (2001) and Teotia (2000). There is a separate chapter on multifractals by Teotia and Kumar (2013).

Geophysical Significance of Wavelets and Fractals

Wavelets in Geophysics

Wavelet transformation has evolved due to the insufficient resolution that FT and WFT offered in determining the time–frequency localization in nonstationary signals.* Because the geophysical signals are, by nature, non-stationary, wavelets play a key role in understanding and interpreting the geophysical data more meaningfully. Jean Morlet, a French geophysicist, first noticed the importance of wavelets in the improved understanding of seismic signals when he provided a better interpretation of the subsur-face geology by analyzing the backscattered seismic signals using wavelets

---

* These are the signals whose statistical properties, such as mean, standard deviation, and others, are time variant.

(Morlet 1981). Later, a thorough mathematical formalism for CWT was made (Goupillaud et al. 1984; Grossmann and Morlet 1984; Grossmann et al. 1989). Recognizing the important limited spatial support that wavelets offer, Farge (1992) employed wavelets to analyze turbulent flows. Wavelet studies of flow patterns later found many applications in solving problems of atmospheric turbulence (Katul et al. 1994; Narasimha 2007, to cite a few), climate studies (Lau and Weng 1995; Torrence and Compo 1998), and space–time rainfall studies (Narasimha and Kailas 2001; Labat et al. 2005; Bhattacharya and Narasimha 2007; Azad et al. 2008). Suitable applications have also been found in geomagnetism. Alexandrescu et al. (1995) and Adhikari et al. (2009) studied the sudden jumps (called geomagnetic jerks) observed in decadal variations of geomagnetic fields using wavelet analysis, as wavelet analysis is best suited for identifying the discontinuities or abrupt changes in the signals. Geomagnetic jerks, which usually have a time span of 1 year, arise due to the combination of a steady flow and a simple time-varying, axisymmetric, equatorially symmetric, toroidal zonal flow of the core fluid (Bloxham et al. 2002). Studies of geomagnetic jerks are important to understand lower mantle conductivity (Alexandrescu et al. 1999). A detailed study on the phase characteristics of geomagnetic jerks using complex wavelets is provided in Chapter 7 of this book (Chandrasekhar et al. 2013). More recently, Kunagu et al. (2013), while studying the characterization of external source fields, applied CWT to 10 years of CHAMP satellite magnetic data and identified some unmodeled signals in geomagnetic field models. Kunagu and Chandrasekhar (2013) later implemented the results of Kunagu et al. (2013) for geomagnetic induction studies. Moreau et al. (1997, 1999), Sailhac et al. (2000, 2009), and Sailhac and Gibert (2003) employed CWT to interpret the geophysical potential field data and to identify the sources of potential fields. Wavelet analysis also found important applications in the analysis and comprehensive understanding of geophysical well-log data, and provided space localizations of different subsurface formations. Wavelet analysis helped to estimate the preferential flow paths and the existence of flow barriers within the reservoir rocks (Jansen and Kelkar 1997), to determine high-frequency sedimentary cycles of oil source rocks (Prokoph and Agterberg 2000), to identify reservoir anomalies from pressure transient data (Panda et al. 2000; Soliman et al. 2001), for detection of cyclic patterns in well-log data (Rivera et al. 2004) and to the gamma ray log data to identify depths to the top of the formation zones (Choudhury et al. 2007). By making a histogram analysis of wavelet coefficients, Chandrasekhar and Rao (2012) optimized the suitable wavelet for studying the geophysical well-log data.

In this book, various other applications and developments in wavelet-based and fractal-based data analysis techniques in various processes of Earth system sciences, namely, paleoclimate studies, a study of non-stationary behaviour of rainfall anomalies using tree-ring data, phase field modeling studies and seismology and earthquake prediction studies using multifractal analyses, etc., are detailed in respective chapters.

Fractals in Geophysics

Various applications of fractals, particularly in the field of earth sciences, are given in the books by Mandelbrot (1983), Turcottee (1992), Feder (1988), Dimri (2000a, 2005a), Dimri et al. (2012), and Srivastava (2013; this volume). A collection of articles on geophysical applications is edited by Scholz and Mandelbrot (1989). Also, the application of fractals in a potential field survey design is given in Dimri (1998) and Srivastava et al. (2007). An application-oriented text on fractal application in reservoir engineering is given by Hardy and Beier (1994) and Dimri et al. (2012). Fractal theory has been extensively used in exploration geophysics, particularly in the interpretation of gravity and magnetic data (Pilkington and Todoeschuck 1993; Maus and Dimri 1994, 1995a,b, 1996; Bansal and Dimri 1999, 2005; Bansal et al. 2006; Fedi et al. 1997; Dimri and Srivastava 2005; Srivastava et al. 2009), and in seismic data (Tommy et al. 2001; Srivastava and Sen 2009, 2010). The application of monofractals and multifractals has been demonstrated in earthquake and tsunami studies [Teotia 2000; Teotia and Kumar 2013 (this volume); Sunmonu and Dimri 1999, 2000; Sunmonu et al. 2001; Mandal et al. 2005; Dimri 2005b; Dimri and Srivastava 2007]. Fractal theory can be applied in soil properties (Dimri 2000b; Ahmadi et al. 2011), biodiversity (Dimri and Ravi Prakash 2001), and chaotic studies (Dimri et al. 2011). Attempts have also been made to understand crustal heat production in the thermal regime of the continents (Dimri and Vedanti 2005; Vedanti et al. 2011). Those interested in the mathematical development of fractal theory are encouraged to follow Falconer (1990).

Geophysical data are often nonstationary or time-varying. There are various methods to convert the time-varying series into piecewise (time invariant) stationary series (Wang 1969; Dimri 1986; Bansal and Dimri 1999), as well as other methods such as Kalman filtering and adaptive approaches (Dimri 1992). Dimri and Srivatsava (1987, 1990) used such an approach for seismic deconvolution problems. After the discovery of wavelets, a combined approach of wavelets and fractals for realistic geology studies was found to be very useful in a series of articles by Dimri et al. (2005), Chamoli and Dimri (2007), and Chamoli et al. (2006, 2007, 2010, 2011).

Relation between Wavelets and Fractals

By now, we understand the concepts of wavelets and fractals. As seen in the previous section, there are several methods to determine the fractal dimension (see Wang et al. 2003 for a review). In addition to the above methods, spectral methods (i.e., Fourier and wavelet methods) are also used to estimate the fractal dimension. McCoy and Walden (1996) and Wang et al. (2003) show that among the spectral methods, the wavelet method gives a better estimate of fractal dimension than the Fourier method. In this section, we further discuss the relation between wavelets and fractals and the advantage of wavelet-based fractal analysis of nonstationary geophysical signals.

Let us consider the example of a Haar wavelet. Figure 1.6 explains that the scaling function, $\varphi(t)$, of the Haar wavelet can be expressed as a sum of its dilated and translated versions, for example, $\varphi(2t)$ (Figure 1.6a) and $\varphi(2t - 1)$ (Figure 1.6b). Mathematically, it is expressed as

$$\varphi(t) = \varphi(2t) + \varphi(2t - 1) \tag{1.21}$$

which gives the scaling function for a Haar wavelet (Figure 1.6c).

A close observation of these scaling functions clearly explains the self-similar nature of the Haar wavelet at different scales, which can be effectively employed for studying the self-similar nature of fractal images using the Haar wavelet. Wavelet transformation proves to be a very efficient tool for analyzing fractal objects. Chapter 4 of this book (and references therein) offers more details of such a description of multiresolution wavelet analysis of fractal images. Using the real Gaussian wavelet, Arneodo et al. (1988) computed the wavelet transform of the standard triadic Cantor set and showed that the wavelet transform allows us to capture the full complexity of the self-similar properties of multifractals. They also claim that the application of the wavelet transform to a variety of physical situations such as percolation, growth phenomena, and fully developed turbulence to name a few, looks very promising. Of late, the use of wavelets in analyzing fractal imaging has gained unprecedented importance. Davis (1998) provides a detailed explanation on the advantages of wavelet based analysis of fractal image compression. By comparing the conventional fractal coding algorithm based on an iterated function system (IFS) employed on the famous "Lenna" and "Baboon" images, and by developing a multiresolution fractal coding algorithm to analyze the above images using the Daubechies wavelet of order 2 (db2), Cesbron and Malassenet (1997) demonstrate the advantage of the latter technique in image compression methods. Recently, Lopez and Aldana (2007) adopted wavelet-based fractal analysis and waveform classifiers to determine fractal parameters from geophysical well-log data.

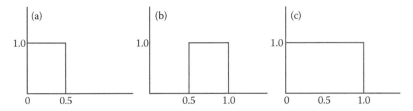

FIGURE 1.6
The scaling functions of Haar wavelet functions: (a) $\varphi(2t)$, (b) $\varphi(2t - 1)$, and (c) their sum, $\varphi(t)$, representing the self-similar nature of the wavelet at different translations and dilations.

Methodology for Wavelet-Based Fractal Analysis

For wavelet-based fractal analysis, first the variance of the wavelet coefficients is obtained at each level of decomposition. Then the log-variance is plotted against the scale. The linear portions in such plots correspond to a power–law process over a particular frequency region, indicating the fractal nature of the phenomena in that frequency range. The slope of the line can be related with the exponent of the power–law process (Percival and Guttorp 1994). Akay (1995) has employed the above methodology and determined the dimension of the echocardiac signals. Using the log-variance of the wavelet coefficients obtained using db2 wavelet (which has self-similar characteristics), Huang and Morimoto (2008) determined the fractal dimension (also known as Hurst exponent) for the El Niño/La Niña episodes. In wavelet-based fractal analysis, because the relation between fractal dimension and scale is established, the fractal dimension obtained through this method, will in fact characterize the signal (a well-log signal or a seismic signal) that could be associated with different sedimentary formations. Enescu et al. (2004) studied the multifractal and correlation properties of real and simulated time series of earthquakes, using a wavelet-based multifractal analysis and quantitatively described the complex temporal patterns of seismicity, their multifractal and clustering properties in particular. Further research in various fields of science and engineering are actively progressing on the application of wavelet-based fractal analysis of nonlinear signals with wavelet coefficients obtained by two-dimensional wavelet transformation.

Conclusions

In this chapter, we have explained the fundamental concepts of wavelets and fractals and their advantages over conventional signal processing techniques in various problems from different fields of science and engineering research. We have also explained (by describing simple examples) and showed the need for these novel wavelet and fractal signal analysis tools to understand these signals in a better and wider perspective. Highlighting the role of wavelets, fractals, and multifractals—particularly in geosciences research—we have provided a number of examples in which these novel signal analysis tools have proven to be very effective. While discussing the relation between wavelets and fractals, we have highlighted the role of wavelets in fractal analysis of geophysical data as well as in image compression techniques. The reader is expected to read all the chapters of this volume and some more books and other literature to have a broader understanding of the subject for its further use in the analysis of a variety of nonstationary signals.

Acknowledgments

The authors thank two anonymous referees for their meticulous reviews, which have improved the quality of this chapter.

References

Adhikari, S., Chandrasekhar, E., Rao, V.E., and Pandey, V.K. 2009. On the wavelet analysis of geomagnetic jerks of Alibag Magnetic Observatory data, India. *Proceedings of the XIII IAGA Workshop on Geomagnetic Observatory Instruments, Data acquisition and Processing*, edited by Jeffery J. Love, 14–23. USGS Open-File Report 2009–1226.

Ahmadi, A., Neyshabouri, M.R., Rouhipour, H., and Asadi, H. 2011. Fractal dimension of soil aggregates as an index of soil erodibility. *Journal of Hydrology* 400, 305–311.

Akay, M. 1995. Wavelets in biomedical engineering. *Annals of Biomedical Engineering* 23(5), 531–542.

Alexandrescu, M., Gibert, D., Hulot, G., Le Mouël, J.L., and Saracco, G. 1995. Detection of geomagnetic jerks using wavelet analysis. *Journal of Geophysical Research* 100, 12557–12572.

Alexandrescu, M., Gibert, D., Le Mouël, J.L., Hulot, H., and Saracco, G. 1999. An estimate of average lower mantle conductivity by wavelet analysis of geomagnetic jerks. *Journal of Geophysical Research* 104, 17735–17745.

Arneodo, A., Grasseau, G., and Holschneider, M. 1988. Wavelet transform of multifractals. *Physical Review Letters* 61, 2281–2284.

Azad, S., Narasimha, R., and Sett, S.K. 2008. A wavelet based significance test for periodicities in Indian monsoon rainfall. *International Journal of Wavelets, Multiresolution and Information Processing* 6, 291–304.

Banks, R. 1975. Complex demodulation of geomagnetic data and the estimation of transfer functions. *Geophysical Journal of the Royal Astronomical Society* 43, 87–101.

Bansal, A.R., and Dimri, V.P. 1999. Gravity evidence for mid crustal structure below Delhi fold belt and Bhilwara super group of western India. *Geophysical Research Letters* 26, 2793–2795.

Bansal, A.R., and Dimri, V.P. 2005. Depth determination from nonstationary magnetic profile for multi scaling geology. *Geophysical Prospecting* 53, 399–410.

Bansal, A.R., Dimri, V.P., and Sagar, G.V. 2006. Depth estimation for gravity data using maximum entropy method (MEM) and multi taper method (MPM). *Pure and Applied Geophysics* 163(7), 1417–1434.

Bhattacharya, S., and Narasimha, R. 2007. Regional differentiation in multidecadal connections between Indian monsoon rainfall and solar activity. *Journal of Geophysical Research* 112, D24103.

Bloxham, J., Zatman, S., and Dumberry, M. 2002. The origin of geomagnetic jerks. *Nature* 420, 65–68.

Cesbron, F.C., and Malassenet, F.J. 1997. Wavelet and fractal transforms for image compression. *Proceedings of the IPA97 Conference No. 443, 15–17 July, 1997*, 77–80.

Chamoli, A., and Dimri, V.P. 2007. Evidence of continental crust over Laxmi Basin (Arabian Sea) using wavelet analysis. *Indian Journal of Marine Sciences* 36(2), 117–121.

Chamoli, A., Srivastava, R.P., and Dimri, V.P. 2006. Source depth characterization of potential field data of Bay of Bengal by continuous wavelet transform. *Indian Journal of Marine Sciences* 35(3), 195–204.

Chamoli, A., Bansal, A.R., and Dimri, V.P. 2007. Wavelet and rescaled range approach for the Hurst coefficient for short and long time series. *Computers and Geosciences* 33, 83–93.

Chamoli, A., Swaroopa Rani, V., Srivastava, K., Srinagesh, D., and Dimri, V.P. 2010. Wavelet analysis of the seismograms for tsunami warning. *Nonlinear Processes in Geophysics* 17(5), 569–574.

Chamoli, A., Pandey, A.K., Dimri, V.P., and Banerjee, P. 2011. Crustal configuration of the northwest Himalaya based on modeling of gravity data. *Pure and Applied Geophysics* 168(5), 827–844.

Chandrasekhar, E. 2000. Geo-electrical structure of the mantle beneath the Indian region derived from the 27-day variation and its harmonics. *Earth, Planets and Space* 52, 587–594.

Chandrasekhar, E., and Rao, V.E. 2012. Wavelet analysis of geophysical well-log data of Bombay offshore basin, India. *Mathematical Geosciences* 44(8), 901–928. doi: 10.1007/s11004-012-9423-4.

Chandrasekhar, E., Prasad, P., and Gurijala, V.G. 2013. Geomagnetic jerks: A study using complex wavelets, In *Wavelets and Fractals in Earth System Sciences*, E. Chandrasekhar et al. (Eds.),195–217. CRC Press, Taylor and Francis, UK.

Choudhury, S., Chandrasekhar, E., Pandey, V.K., and Prasad, M. 2007. Use of wavelet transformation for geophysical well-log data analysis. *IEEE Xplore* 647–650. doi: 10.1109/ICDSP.2007.4288665.

Chui, C.K. 1992. *An Introduction to Wavelets*. Academic Press, New York.

Daubechies, I. 1992. *Ten Lectures on Wavelets*. SIAM, Philadelphia, PA.

Davis, G.M. 1998. A wavelet-based analysis of fractal image compression. *IEEE Transactions on Image Processing*, 7(2), 141–154.

Dimri, V.P. 1986. On the time varying Wiener filter. *Geophysical Prospecting* 34, 904–912.

Dimri, V.P. 1992. *Deconvolution and Inverse Theory: Application to Geophysical Problems*, 230. Elsevier Science Publishers, Amsterdam.

Dimri, V.P. 1998. Fractal behavior and detectibility limits of geophysical surveys. *Geophysics* 63, 1943–1947.

Dimri, V.P., (ed.). 2000a. *Application of Fractals in Earth Sciences*, 238. A.A. Balkema, Rotterdam.

Dimri, V.P. 2000b. Fractal dimension analysis of soil for flow studies. In *Application of Fractals in Earth Sciences*, edited by V.P. Dimri, 189–193. A.A. Balkema, Rotterdam.

Dimri, V.P. 2000c. Crustal fractal magnetization. In *Application of Fractals in Earth Sciences*, edited by V.P. Dimri, 89–95. A.A. Balkema, Rotterdam.

Dimri, V.P., ed. 2005a. *Fractal Behaviour of the Earth System*, 1–22. New York, Springer.

Dimri, V.P. 2005b. Fractals in geophysics and seismology: An introduction. In *Fractal Behaviour of the Earth System*, edited by V.P. Dimri, 207. New York, Springer.

Dimri, V.P., and Srivastava, K. 1987. Ideal performance criteria for deconvolution operator. *Geophysical Prospecting* 35, 539–547.

Dimri, V.P., and Srivastava, K. 1990. The optimum gate length for time varying deconvolution operator. *Geophysical Prospecting* 38, 405–410.

Dimri, V.P., and Srivastava, R.P. 2005. Fractal modeling of complex subsurface geological structures, In *Fractal Behaviour of the Earth System*, edited by V.P. Dimri, 23–37. New York, Springer.

Dimri, V.P., and Srivastava, K. 2007. Tsunami Propagation of the 2004 Sumatra Earthquake and the Fractal Analysis of the Aftershock Activity, *Indian Journal of Marine Sciences* 36(2), 128–135.

Dimri, V.P., and Ravi Prakash, M. 2001. Scaling of power spectrum of extinction events in the fossil record. *Earth and Planetary Science Letters* 186, 363–370.

Dimri, V.P., and Vedanti, N. 2005. Scaling evidences of thermal properties in Earth's crust and its implications. In *Fractal Behaviour of the Earth System*, edited by V.P. Dimri, 119–132. New York, Springer.

Dimri, V.P., Vedanti, N., and Chattopadhyay, S. 2005. Fractal analysis of aftershock sequence of Bhuj earthquake — a wavelet based approach, *Current Science* 88(10), 1617–1620.

Dimri, V.P., Srivastava, R.P., and Vedanti, N. 2011. Fractals and chaos. In *Encyclopedia of Solid Earth Geophysics*, edited by H.K. Gupta, 297–302. Dordrecht, Springer.

Dimri, V.P., Srivastava, R.P., and Vedanti, N. 2012. *A Handbook on 'Fractal Models in Exploration Geophysics.'* Elsevier Science Publishers, Amsterdam.

Enescu, B., Ito, K., and Struzik, Z.R. 2004. Wavelet-based multifractal analysis of real and simulated time series of earthquakes. *Annuals of Disaster Prevention Research Institute, Kyoto University*, 47B, 1–13.

Falconer, K. 1990. *Fractal Geometry: Mathematical Foundations and Applications*. John Wiley & Sons, England.

Farge, M. 1992. Wavelet transforms and their applications to turbulence. *Annual Reviews of Fluid Mechanics* 24, 395–457.

Feder, J. 1988. *Fractals*. Plenum Press, New York.

Fedi, M., Quarta, T., and Sanits, A.D. 1997. Inherent power-law behavior of magnetic field power spectra from a Spector and Grant ensemble. *Geophysics* 62, 1143–1150.

Frisch, U., and Parisi, G. 1985. Fully developed turbulence and intermittency. In *Turbulence and Predictability in Geophysical Fluid Dynamics and Climate Dynamics*, edited by, M. Ghil 84. North Holland, Amsterdam.

Gabor, D. 1946. Theory of communication. *Journal of the Institute of Electrical Engineers* 93, 429–441.

Goupillaud, P., Grossmann, A., and Morlet, J. 1984. Cycle-octave and related transforms in seismic signal analysis. *Geoexploration* 23, 85–105.

Graps, A.L. 1995. An introduction to wavelets. *IEEE Computational Sciences and Engineering* 2, 50–61.

Grassberger, P., Badii, R., and Politi, A. 1988. Scaling laws for invariant measures on hyperbolic attractors. *Journal of Statistical Physics* 51, 135–178.

Grossmann, A., and Morlet, J. 1984. Decomposition of Hardy functions into square integrable wavelets of constant shape. *SIAM Journal on Mathematical Analysis* 15, 723–736.

Grossmann, A., Kronland-Martinet, R., and Morlet, J. 1989. Reading and understanding continuous wavelet transforms. In *Wavelets, Time-Frequency Methods and Phase Space*, 1st International Wavelet ConTr., Marseille, December 1987, Inverse Probl. Theoret. Imaging, edited by, J.M. Combes, A. Grossmann, and P. Tchamitchian, 2–20. New York, Springer.

Hardy, H.H., and Beier, R.A. 1994. *Fractals in Reservoir Engineering*. River Edge, NJ, World Scientific Publishing Company.

Hentshel, H.G.E., and Procaccia, J. 1983. The infinite number of generalized dimensions of fractals and strange attractors. *Physica* 8D, 435–444.

Hirabayashi, T., Ito, K., and Yoshii, T. 1992. Multifractal analysis of earthquakes. *PAGEOPH* 138(4), 591–610.

Hirata, T. 1989. A correlation between b-value and the fractal dimension of the earthquakes. *Journal of Geophysical Research* 94, 7507–7514.

Huang, Z., and Morimoto, H. 2008. Wavelet based fractal analysis of El Nino/La Nina episodes. *Hydrological Research Letters* 2, 70–74.

Jansen, F.E., and Kelkar, M. 1997. Application of wavelets to production data in describing inter-well relationships. Society of Petroleum Engineers, SPE 38876. In *Annual Technical Conference and Exhibition, San Antonio, TX, Oct. 5–8*.

Kaiser, G. 1994. A Friendly Guide to Wavelets. Birkhäuser.

Katul, G.G., Albertson, J.D., Chu, C.R., and Parlange, M.B. 1994. Intermittency in atmospheric surface layer turbulence: The orthonormal wavelet representation. In *Wavelets in Geophysics*, edited by E. Foufoula-Georgiou and P. Kumar, 81–106. New York: Academic Press.

Kunagu, P., and Chandrasekhar, E. 2013. External field characterization using CHAMP satellite data for induction studies. *Journal of Earth System Science* 122(3), 651–660. doi: 10.1007/s12040-013-0306-y.

Kunagu, P., Balasis, G., Lesur, V., Chandrasekhar, E., and Papadimitriou, C. 2013. Wavelet characterization of external magnetic sources as observed by CHAMP satellite: Evidence for unmodelled signals in geomagnetic field models. *Geophysical Journal International* 192, 946–950. doi: 10.1093/gji/ggs093.

Labat, D., Ronchailb, J., and Guyotb, J.L. 2005. Recent advances in wavelet analyses: Part 2—Amazon, Parana, Orinoco and Congo discharges time scale variability. *Journal of Hydrology* 314, 289–311.

Lau, K.-M., and Weng, H. 1995. Climate signal detection using wavelet transform: How to make a time series sing. *Bulletin of the American Meteorological Society* 76, 2391–2402.

Lopez, M., and Aldana, M. 2007. Facies recognition using wavelet based fractal analysis and waveform classifier at the Oritupano-A field, Venezuela. *Nonlinear Processes in Geophysics* 14, 325–335.

Ma, K., and Tang, X. 2001. Translation-invariant face feature estimation using discrete wavelet transform, *Wavelet Analysis and Its Applications*, In Y.Y. Tang et al. (Eds.), LNCS 2251, 200–210. Springer-Verlag, Berlin, Heidelberg.

Malamud, B.D. and Turcotte, D.L. 1999. Self-affine time series: I. Generation and analyses. *Advances in Geophysics* 40, 1–90.

Mallat, S. 1989. A theory for multiresolution signal decomposition: The wavelet representation. *IEEE Transactions on Pattern Analysis and Machine Intelligence* 11, 674–693.

Mallat, S. 1999. *A Wavelet Tour of Signal Processing*. San Diego, CA: Academic Press.

Mandal, P., Mabawonku, A.O., and Dimri, V.P. 2005. Self-organized fractal seismicity of reservoir triggered earthquakes in the Koyna–Warna seismic zone, western India. *Pure and Applied Geophysics* 162, 73–90.

Mandelbrot, B.B. 1967. How long is the coast of Britain? Statistical self-similarity and fractional dimension. *Science* 156, 636–638.

Mandelbrot, B.B. 1983. The fractal geometry of nature. W.H. Freeman & Co, ISBN 0-7167-1186-9.

Maus, S., and Dimri, V.P. 1994. Scaling properties of potential fields due to scaling sources. *Geophysical Research Letters* 21, 891–894.

Maus, S., and Dimri, V.P. 1995a. Potential field power spectrum inversion for scaling geology. *Journal of Geophysical Research* 100, 12605–12616.

Maus, S., and Dimri, V.P. 1995b. Basin depth estimation using scaling properties of potential fields. *Journal of the Association of Exploration Geophysicists* 16,131–139.

Maus, S., and Dimri, V.P. 1996. Depth estimation from the scaling power spectrum of potential field? *Geophysical Journal International* 124, 113–120.

McCoy, E.J., and Walden, A.T. 1996. Wavelet analysis and synthesis of stationary long-memory process. *Journal of Computational and Graphical Statistics* 5, 26–56.

Moharir, P.S. 2000. Multi fractal, Edited: *Application of Fractals in Earth Sciences*, 45–58, A.A. Balkema, Rotterdam.

Moreau, F., Gibert, D., Holschneider, M., and Saracco, G. 1997. Wavelet analysis of potential fields. *Inverse Problems* 13, 165–178.

Moreau, F., Gibert, D., Holschneider, M., and Saracco, G. 1999. Identification of sources of potential fields with the continuous wavelet transform: Basic theory. *Journal of Geophysical Research* 104, 5003–5013.

Morlet, J. 1981. Sampling theory and wave propagation. In *Proceedings of the 51st Annual Meeting of the Society of Exploration Geophysics*, Los Angeles, CA.

Naidu, P.S. 1970. Statistical structure of aeromagnetic field. *Geophysics* 35, 279–292.

Narasimha, R. 2007. Wavelet diagnostics for detection of coherent structures in instantaneous turbulent flow imagery: A review. *Sadhana* 32, 29–42.

Narasimha, R., and Kailas, S.V. 2001. A wavelet map of monsoon variability. *Proceedings of the Indian National Science Academy* 67, 327–341.

Negi, J.G., Dimri, V.P., Agarwal P.K., and Pandey, O.P. 1986. A spectral analysis of the aeromagnetic profiles for thickness estimation of flood basalts of India. *Exploration Geophysics* 17, 105–111.

Panda, M.N., Mosher, C.C., and Chopra, A.K. 2000. Application of wavelet transforms to reservoir-data analysis and scaling. *Society of Petroleum Engineers* 5, 92–101.

Percival, D.B., and Guttorp, P. 1994. Long-memory processes, the Allan variance and wavelets. In *Wavelets in Geophysics*, edited by E. Foufoula-Georgiou, and P. Kumar, 4, 325–344. Academic Press. San Diego, California.

Percival, D.B., and Walden, A.T. 2000. *Wavelet Methods for Time Series Analysis.* Cambridge, UK: Cambridge University Press.

Pilkington, M., and Todoeschuck, J.P. 1993. Fractal magnetization of continental crust. *Geophysical Research Letters* 20, 627–630.

Polikar, R. 2001. The engineer's ultimate guide to wavelet analysis: The wavelet tutorial. http://users.rowan.edu/~polikar/wavelets/wttutorial.html.

Prokoph, A., and Agterberg, F.P. 2000. Wavelet analysis of well-logging data from oil source rock, Egret Member, offshore Eastern Canada. *Bulletin of American Association of Petroleum Geologists* 84, 1617–1632.

Rangarajan, G. 2000. Fractals, Edited: *Application of Fractals in Earth Sciences*, 7–16. A.A. Balkema, Rotterdam.

Rivera, N., Ray, S., Jensen, J., Chan, A.K., and Ayers, W.B. 2004. Detection of cyclic patterns using wavelets: An example study in the Ormskirk sandstone, Irish Sea. *Mathematical Geology* 36, 529–543.

Sailhac, P., and Gibert, D. 2003. Identification of sources of potential fields with the continuous wavelet transform: Two-dimensional wavelets and multipolar approximations. *Journal of Geophysical Research* 108, 2296–2306.

Sailhac, P., Galdeano, A., Gibert, D., Moreau, F., and Delor, C. 2000. Identification of sources of potential fields with the continuous wavelet transform: Complex wavelets and application to aeromagnetic profiles in French Guiana. *Journal of Geophysical Research* 105, 19,455–19,475.

Sailhac, P., Gibert, D., and Boukerbout, H. 2009. The theory of the continuous wavelet transform in the interpretation of potential fields: a review. *Geophysical Prospecting* 57, 517–525.

Scholz, C.H., and Mandlebrot, B.B. (eds.) 1989. *Fractals in Geophysics*. Boston, Birkhauser.

Sharma, A., Bunde, A. Dimri, V.P., and Baker, D.N. (eds.). 2012. Extreme events and natural hazards: The complexity perspective. *Geophysical Monograph Series* 196, 371.

Soliman, M.Y., Ansah, J., Stephenson, S., and Manda, B. 2001. Application of wavelet transform to analysis of pressure transient data. Society of Petroleum Engineers, Abstract no. SPE 71571. In *Annual Technical Conference and Exhibition, New Orleans, September 30–October 3, 2001*.

Spector, A., and Grant, F.S. 1970. Statistical model for interpreting aeromagnetic data. *Geophysics* 35(2), 293–302.

Srivastava, R.P. 2013. Fractals and wavelets in applied geophysics with some examples, In *Wavelets and Fractals in Earth System Sciences*, E. Chandrasekhar et al. (Eds.) 155–176. CRC Press, Taylor and Francis, UK.

Srivastava, R.P., and Sen, M.K. 2009. Fractal based stochastic inversion of post stack seismic data using very fast simulated annealing. *Journal of Geophysics and Engineering* 6, 412–425.

Srivastava, R.P., and Sen, M.K. 2010. Stochastic inversion of prestack seismic data using fractal based initial models. *Geophysics* 75(3), R47–R59.

Srivastava, R.P., Vedanti, N., and Dimri, V.P. 2007. Optimum design of a gravity survey network and its application to delineate the Jabera–Damoh structure in the Vindhyan Basin, Central India. *Pure and Applied Geophysics* 164, 1–14.

Srivastava, R.P., Vedanti, N., Pandey, O.P., and Dimri, V.P. 2009. Detailed gravity studies over Jabera–Damoh region of the Vindhyan Basin (Central India) and crustal evolution. *Journal of the Geological Society of India* 73, 715–723.

Sunmonu, L.A., and Dimri, V.P. 1999. Fractal analysis and seismicity of Bengal basin and Tripura fold belt, Northeast India. *JGSI* 53, 587–592.

Sunmonu, L.A., and Dimri, V.P. 2000. Fractal geometry of faults and seismicity of Koyna–Warna region west India using Landsat images. *Pure and Applied Geophysics* 157, 1393–1405.

Sunmonu, L.A., Dimri, V.P., Ravi Prakash, M., and Bansal, A.R. 2001. Multifractal approach of the time series of M > 7 earthquakes in Himalayan region and its vicinity during 1985–1995. *JGSI* 58, 163–169.

Teotia, S.S. 2000. Multifractal analysis of earthquakes: An overview, edited by V.P. Dimri, *Application of Fractals in Earth Sciences* 161–170. A.A. Balkema, Rotterdam.

Teotia, S.S., and Kumar, D. 2013. Role of multifractal study in earthquake prediction, In *Wavelets and Fractals in Earth System Sciences*, E. Chandrasekhar et al. (Eds.), 177–194. CRC Press, Taylor and Francis, UK.

Tommy, T., Dimri, V.P., and Ursin, B. 2001. Comparison of deconvolution methods for scaling reflectivity. *Journal of Geophysics* XXII, 2–4, 117–123.

Torrence, C., and Compo, G.P. 1998. A practical guide to wavelet analysis. *Bulletin of the American Meteorological Society* 79, 61–78.

Turcottee, D.L. 1992. *Fractals and Chaos in Geology and Geophysics*. Cambridge University Press, New York.

Vedanti, N., and Dimri, V.P. 2003. Fractal behavior of electrical properties in oceanic and continental crust. *Indian Journal of Marine Sciences* 32(4), 273–278.

Vedanti, N., Srivastava, R.P., Pandey, O.P., and Dimri, V.P. 2011. Fractal behaviour in continental crustal heat production. *Nonlinear Processes in Geophysics* 18, 119–124.

Wang, R.J. 1969. The determination of optimum gate lengths for time varying Wiener filtering. *Geophysics* 34, 683–695.

Wang, J.H. and Lee, CW. 1996, Multifractal measures of earthquakes in west Taiwan. *Pure and Applied Geophysics* 196, 131–145.

Wang, A.L., Yang, C.X., and Yuan, X.G. 2003. Evaluation of the wavelet transform method for machined surface topography: Methodology validation. *Tribology International* 36, 517–526.

Zhang, Y., and Paulson, K.V. 1997. Enhancement of signal-to-noise ratio in natural-source transient magnetotelluric data with wavelet transform. *Pure and Applied Geophysics* 149, 409–415.

2

Construction of Wavelets: Principles and Practices

Manish Sharma, Ashish V. Vanmali, and Vikram M. Gadre

CONTENTS

Introduction 30
Piecewise Constant Approximation and Haar Wavelet 32
Perfect Reconstruction Filter Banks..38
    Analysis and Synthesis Filter Banks..38
    Two-Channel PR Filter Banks..40
    Product Filter and Design of PR Filter Banks.....................................43
        Half-Band Filter 44
    Paraunitary Filter Banks...44
    Linear Phase Biorthogonal Filter Banks..47
    Maxflat Half-Band Filter 48
Octave Filter Banks 51
CWT versus DWT 52
Connection between Filter Banks and Wavelets..54
    Multiresolution Analysis..54
    Biorthogonal MRA...55
    Fast Wavelet Transform 57
    Wavelet Filter Banks: The Special Choice of PR Filter Banks 59
    Generating Scaling Functions and Wavelets: The Cascade Algorithm 60
    A Necessary Condition for the Convergence in $L_2(\mathbb{R})$ 63
        Transition Matrix 64
    Measure of Regularity 68
Design of Wavelet Filter Banks 69
Design of Orthogonal Wavelets 71
    Design of Compactly Supported Orthogonal Wavelets on the
    Interval [0, 1] 72
    Design of Compactly Supported Orthogonal Wavelets on the
    Interval [0, 3] 72
Design of Linear Phase Biorthogonal Filter Banks: Parametrization
Technique 73
Time and Frequency Conflict: The Uncertainty Principle............................76
    Uncertainty Principle: Continuous-Time Signal................................76

Uncertainty Principle: Discrete-Time Signal 78
   A Direct Extension from the Continuous-Time Measure 78
Filter Bank Design: Time–Frequency Localization..79
Our Design Methodology 80
   Results and Comparisons..81
Conclusions..90
References..90

Introduction

The aim of signal analysis is to achieve a better understanding and description of the underlying physical phenomenon and to find a compact representation of signals, which leads to compact storage and efficient transmission of signals. A wide variety of transforms or operators are available for analyzing signals, which range from the classical Fourier transform (FT) to various types of joint time-frequency based operators such as the short-time Fourier transform (STFT), Wigner–Ville distribution (WVD), block transform (BT), and wavelet transform (WT). The Fourier transform brings forth many useful features of a signal and hence, transforms that give the spectral information of a signal have been used extensively. For stationary or wide-sense stationary and time-invariant phenomena, the Fourier transform is sufficient to characterize the spectral distribution of energy because this distribution does not change with time. However, for transient, nonstationary, or time-varying phenomena in which the spectral distribution changes with time, Fourier transform is not suitable because it would sweep the entire time axis and wash out any local features (e.g., bursts of high frequency) in the signal. Useful information is contained in the nonglobal features of a signal in many situations. Hence, this major drawback of traditional Fourier analysis stimulates us to search for transforms that would give us a better view of the local properties of a signal. To overcome this, Gabor introduced the STFT, which calculates the windowed Fourier transform by moving a window over the time axis. Thus, we extract the frequency information in the interval of the window only. In this transform, the inner product of the signal with a compactly supported modulated window is calculated. The window is then shifted in time and the inner product is calculated again. This gives us a way to look into the spectral properties of the function around locations determined by the shift parameter. The STFT is a time–frequency representation, that is, it is a function of two variables, frequency and shift parameter. The STFT cannot be defined uniquely unless we specify the window. The choice of window is an important issue, and calls for a trade-off between time localization and frequency resolution.

For the analysis of nonstationary signals, the WT has been proved to be a very efficient and popular tool over the past few decades. To calculate the WT, we shift and scale the window function instead of shifting and modulating a window function as in the case of the STFT. The word wavelet stands for "small wave" or, more technically, a band-pass function with zero mean. The WT essentially expands the signal in terms of wavelet functions, which are localized in both time and frequency unlike the Fourier transform, which expands the signal in terms of everlasting sinusoids. Fourier analysis is global in the sense that each frequency (time) component of the signal is affected by all time (frequency) component of the signal. On the other hand, wavelet analysis is a form of local analysis. This local nature of wavelet analysis makes it suitable for the time–frequency analysis of signals.

Both the STFT and the WT would inherently be functions of two variables: one specifying the location in time and the other specifying the location in frequency. However, varying the two parameters in a continuous fashion results in a lot of redundancy on account of a one-dimensional continuous function being represented in terms of a two-dimensional continuous function. We can get rid of the redundancy by discretizing the values of the two parameters in a certain manner, ensuring that no loss of information occurs. In the case of the WT, the discretization of parameters brought to light the connection between the WT and multirate filter banks along with the notion of multiresolution signal decomposition. These two formulations have been developed independently with different perspectives. Grossman and Morlet introduced the mathematical aspects of the WT. The fundamental contributions by Mallat and Daubechies accelerated research in the area of wavelets. Daubechies constructed compactly supported wavelets and established a connection with filter bank methods. Filter bank methods had already been employed in speech processing applications even before wavelets were formally introduced.

The WT is often preferred over the STFT in certain applications due to the fact that unlike the STFT, which has a fixed resolution in both domains, the WT allows us to improve the resolution in one domain at the expense of that in the other domain. This is often useful, as we can obtain good time resolution for high-frequency effects and good frequency resolution for low-frequency effects, which is well suited for analysis.

Many signals in real life such as speech, audio, and biomedical signals have a nonstationary local spectrum. Some applications such as transient location require knowledge of the characteristics of the signal not only in time or in frequency but also jointly in the time–frequency plane. In such cases, good joint time–frequency localization of the analyzing window is desirable. However, the resolution in both time and frequency cannot be made arbitrarily small because of the lower bound posed by Heisenberg's uncertainty principle. Designing analyzing functions or bases that have good joint time–frequency resolution is a theme that still offers opportunities for exploration. We touch on our work related to this problem in a later part of this chapter.

In summary, in this chapter, we review the connection between filter banks and wavelets, and describe our efforts to design biorthogonal wavelets with the objective of optimizing time localization and frequency resolution simultaneously toward the end. Our aim is to represent various existing notions of time–frequency localization associated with continuous functions and discrete sequences, and how we can quantify (measure) time–frequency localization of wavelet filter banks and corresponding wavelets. Certain important conditions for the convergence of the cascade algorithm are stated as part of the design description at appropriate points in the chapter.

Piecewise Constant Approximation and Haar Wavelet

The process of signal analysis or expansion, using wavelets, has a nice property called *successive approximation,* that is, synthesis or reconstruction of the signal using only an appropriate subset of the basis functions often yields a good approximation of the signal. Furthermore, more bases can be added progressively into the subset of basis functions to have more accurate approximation. This ability to decompose signals into several levels of resolutions is one of the attractive features of wavelet analysis. Therefore, wavelet-based expansion has wide applications in image compression and transmission. In this section, we discuss the piecewise constant approximation property of the Haar wavelet system. We also illustrate how one can construct the Haar wavelet and corresponding scaling function, from the discrete-time filter sequences of the underlying Haar filter bank.

The signal $x(t)$ in Figure 2.1 is divided into small segments spanning the time duration T. We represent or approximate each segment with its piecewise constant approximation, which is the mean value of the signal in the

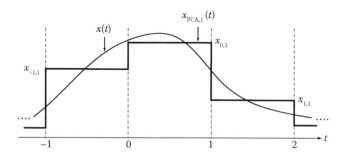

FIGURE 2.1
Piecewise constant approximation on unit interval.

segment. The piecewise constant approximation x_T or average value over an interval T, of a signal $x(t)$, can be computed as

$$x_T = \frac{1}{T} \int_T x(t)\,dt \tag{2.1}$$

We can compute the average value, $x_{n,1}$, of the signal over the standard unit interval $]n, n+1[$, $n \in \mathbb{Z}$, as

$$x_{n,1} = \int_n^{n+1} x(t)\,dt \tag{2.2}$$

Note that in the notation $x_{n,1}$, the first subscript indicates the beginning of the interval and the second subscript indicates the length of the time interval of the segment.

Figure 2.1 depicts the original signal $x(t)$ and its piecewise contact approximation $x_{\mathrm{PCA},1}(t)$. For the notation $x_{\mathrm{PCA},T}(t)$, the subscript PCA is an acronym for piecewise constant approximation, and the subscript T represents the duration of the time interval of each segment of the signal. In this case, the time interval is the standard unit interval.

The box function $\phi(t)$ is defined over a unit interval, as shown in Figure 2.2. Piecewise constant approximation $x_{\mathrm{PCA},1}(t)$, of the signal $x(t)$ over the unit interval, can be expressed as a weighted sum of shifted versions of the box function $\phi(t)$ in the following way:

$$x_{\mathrm{PCA},1}(t) = \ldots + x_{-1,1}\phi(t+1) + x_{0,1}\phi(t) + x_{1,1}\phi(t-1) + \ldots$$

$$= \sum_n x_{n,1}\phi(t-n) \tag{2.3}$$

Here, weights $x_{n,1}$ are average values of the signal taken in the respective unit interval $]n, n+1[$, $n \in \mathbb{Z}$.

FIGURE 2.2
Box function $\phi(t)$.

In general, the approximation signal $x_{PCA,T}(t)$ is not an exact replica of the signal $x(t)$. However, the error $x(t) - x_{PCA,T}(t)$, between the signal $x(t)$ and the piecewise constant approximation $x_{PCA,T}(t)$, can be made arbitrarily small by reducing the duration T of the segment. To reduce the error, the duration of the segment is reduced to ½ and the signal $x(t)$ is now approximated by a piecewise constant approximation taken over intervals of ½. Figure 2.3 illustrates this piecewise constant approximation.

Here, the average $x_{n,\frac{1}{2}}$ value of the signal over the standard half interval $\left] \dfrac{n}{2}, \dfrac{n+1}{2} \right[, n \in \mathbb{Z}$ is computed as

$$x_{n,\frac{1}{2}} = \frac{1}{\frac{1}{2}} \int_{n/2}^{(n+1)/2} x(t)\, dt \tag{2.4}$$

The piecewise constant approximation, $x_{PCA,\frac{1}{2}}(t)$, of the signal, $x(t)$, over the standard half intervals can be expressed as

$$x_{PCA,\frac{1}{2}}(t) = \sum_{n} x_{n,\frac{1}{2}} \phi(2t - n) \tag{2.5}$$

In this case, the approximation error will be less than that in the case of approximation with a unit interval. We can continue this process by making the interval smaller and smaller. Thus, we can approach $x(t)$ to an arbitrary degree of "closeness." On the contrary, if we increase the approximation interval, it will create more and more approximation error. In general, for the

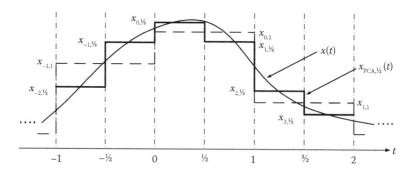

FIGURE 2.3
Piecewise constant approximation on an interval of ½.

interval $]n2^{-m}, (n + 1)2^{-m}[; n, m \in \mathbb{Z}$, of the duration 2^{-m}, the piecewise constant approximation $x_{\text{PCA},2^{-m}}(t)$, of the signal $x(t)$, can be expressed as

$$x_{\text{PCA},2^{-m}}(t) = \sum_{n} x_{n,2^{-m}} \phi(2^m t - n) \tag{2.6}$$

The piecewise constant approximation $x_{\text{PCA},\frac{1}{2}}(t)$ clearly furnishes more information than $x_{\text{PCA},1}(t)$ about the signal $x(t)$. This can be seen graphically in Figure 2.3. The "additional information" provided by $x_{\text{PCA},\frac{1}{2}}(t)$ over $x_{\text{PCA},1}(t)$ is

$$\Delta x_{\frac{1}{2}}(t) = x_{\text{PCA},\frac{1}{2}}(t) - x_{\text{PCA},1}(t) \tag{2.7}$$

Figure 2.4 illustrates this additional information signal $\Delta x_{\frac{1}{2}}(t)$. Let us consider the function $\psi(t)$ as shown in Figure 2.5. We can express the additional information $\Delta x_{\frac{1}{2}}(t)$ as a weighted summation of the shifted versions of $\psi(t)$ as

$$\Delta x_{\frac{1}{2}}(t) = \ldots + (-\Delta x_{-1,1})\psi(t+1) + \Delta x_{0,1}\psi(t) + \Delta x_{1,1}\psi(t-1) + \ldots$$

$$= \sum_{n} \Delta x_{n,1}\psi(t-n) \tag{2.8}$$

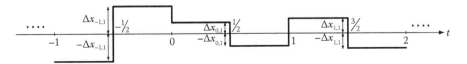

FIGURE 2.4
Additional information captured by an interval of ½.

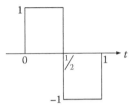

FIGURE 2.5
Function $\psi(t)$.

In particular, the additional information captured, as we move from an interval of $2^{-(m+1)}$ to one of 2^{-m}, can be expressed as

$$\Delta x_{2^{-(m+1)}}(t) = \sum_n \Delta x_{n,2^{-m}} \psi(2^{-m} t - n) \tag{2.9}$$

Equations 2.6 and 2.9 give an important insight. Using a single function, $\phi(t)$, along with its dilated and translated versions, we can construct a piecewise constant approximation over any interval 2^{-m}. Also, using a single function $\psi(t)$, along with its dilated and translated versions, we can construct additional information going from an interval of $2^{-(m+1)}$ to one of 2^{-m}. This is the central idea of Haar multiresolution analysis (MRA). In particular, the function $\phi(t)$ shown in Figure 2.2 is called the *Haar scaling function* and the function, $\psi(t)$, shown in Figure 2.5, is called the *Haar wavelet function*.

From Figure 2.3, it can be observed that $x_{n,1}$ and $\Delta x_{n,1}$ can be obtained from $x_{2n,\frac{1}{2}}$ and $x_{2n+1,\frac{1}{2}}$ using equations

$$x_{n,1} = \frac{x_{2n,\frac{1}{2}} + x_{2n+1,\frac{1}{2}}}{2} \tag{2.10}$$

and

$$\Delta x_{n,1} = \frac{x_{2n,\frac{1}{2}} - x_{2n+1,\frac{1}{2}}}{2} \tag{2.11}$$

These equations are called *analysis equations* and are used to construct the analysis filter bank for the Haar wavelet.

On the other hand, $x_{2n,\frac{1}{2}}$ and $x_{2n+1,\frac{1}{2}}$, can be calculated from $x_{n,1}$ and $\Delta x_{n,1}$ by simple adjustment of Equations 2.10 and 2.11 as follows:

$$x_{2n,\frac{1}{2}} = x_{n,1} + \Delta x_{n,1} \tag{2.12}$$

and

$$x_{2n+1,\frac{1}{2}} = x_{n,1} - \Delta x_{n,1} \tag{2.13}$$

These equations are called *synthesis equations* and are used to construct the synthesis filter bank for the Haar wavelet.

The factor of $\frac{1}{2}$ present in Equations 2.10 and 2.11 can be spread equally on the analysis and the synthesis sides and in that case Equations 2.10, 2.11, 2.12, and 2.13 can be rewritten as

$$x_{n,1} = \frac{1}{\sqrt{2}} \left\{ x_{2n,\frac{1}{2}} + x_{2n+1,\frac{1}{2}} \right\} \tag{2.14}$$

$$\Delta x_{n,1} = \frac{1}{\sqrt{2}} \left\{ x_{2n,\frac{1}{2}} - x_{2n+1,\frac{1}{2}} \right\} \tag{2.15}$$

and

$$x_{2n,\frac{1}{2}} = \frac{1}{\sqrt{2}} \left\{ x_{n,1} + \Delta x_{n,1} \right\} \tag{2.16}$$

$$x_{2n+1,\frac{1}{2}} = \frac{1}{\sqrt{2}} \left\{ x_{n,1} - \Delta x_{n,1} \right\} \tag{2.17}$$

The above equations can be implemented as a filter bank as shown in Figure 2.6. The analysis Equations 2.14 and 2.15 indicate that piecewise constant approximations over a unit interval and the incremental information are obtained as weighted sums and differences of piecewise constant approximations over half-intervals, respectively, with a weight of $1/\sqrt{2}$. This is implemented by filters $H_0(z)$ and $H_1(z)$ in the analysis filter bank. Also, two samples of piecewise constant approximations over half-intervals contribute to a single sample of piecewise constant approximation over a unit interval and the incremental information. This can be achieved by downsampling the outputs of $H_0(z)$ and $H_1(z)$ by 2. On the synthesis side, upsampling by 2 will outdo downsampling on the analysis side. The sequences obtained after upsampling are added and subtracted alternately to recover the original sequence. These operations performed by filters $F_0(z)$ and $F_1(z)$ are shown in Figure 2.6.

The piecewise constant approximation is not the only choice of approximation. One can select different compactly supported approximation functions and different design procedures, leading to different families of wavelets. In general, different design procedures result in different filters $H_0(z)$, $H_1(z)$, $F_0(z)$, and $F_1(z)$. The details of the design procedure are presented in subsequent sections.

For complete details about piecewise constant approximation, readers are referred to Gadre [1].

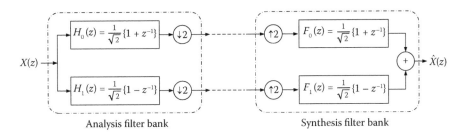

Analysis filter bank   Synthesis filter bank

FIGURE 2.6
Haar filter bank.

Perfect Reconstruction Filter Banks

In a previous section (Piecewise Constant Approximation and Haar Wavelet), we summarized the close connection between MRA and Haar filter banks. In this section, we move from the specific case of the Haar filter bank toward more generalized wavelet filter banks. We obtain certain conditions on two-channel filter banks, for perfect reconstruction leading to the design of perfect reconstruction (PR) two-channel filter banks.

Analysis and Synthesis Filter Banks

A filter bank is an array of low-pass, band-pass, and high-pass filters linked by downsamplers (decimators) and upsamplers (interpolators) with a common input or a common output.

The decimator and interpolator are shown in Figure 2.7. The output of an M-fold decimator (downsampler) retains only those input samples that are multiples of M, that is,

$$y(n) = x(Mn) \tag{2.18}$$

Example:

Figure 2.8 depicts an example where the input sequence $x(n)$ is down-sampled by a factor of two, which yields decimated output sequence $y(n)$.

The input and output of an L-fold interpolator (upsampler) are related as

$$y(n) = \begin{cases} x\left(\dfrac{n}{L}\right) & n = \text{multiple of } L \\ 0 & \text{otherwise} \end{cases} \tag{2.19}$$

Example:

Figure 2.9 shows an example where the input sequence $x(n)$ is upsampled by a factor of two to produce the interpolated sequence $y(n)$. Equation 2.19 implies that the output is obtained by inserting $L - 1$ zeros between two consecutive samples of the input for an upsampling factor of L.

M-fold decimator   L-fold interpolator

FIGURE 2.7
Decimator and interpolator.

FIGURE 2.8
Downsampling by a factor of 2.

FIGURE 2.9
Upsampling by a factor of 2.

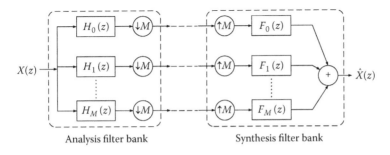

Analysis filter bank   Synthesis filter bank

FIGURE 2.10
M-channel maximally decimated filter bank.

We can classify filter banks into two types: the analysis bank and the synthesis bank. An analysis bank is a collection of filters $\{H_k(z), 0 \le k \le M - 1\}$, which splits the input signal $x(n)$ into M components called subbands of the original signal. A synthesis bank is an array of filters $\{F_k(z), 0 \le k \le M - 1\}$, which combines subband signals, obtained from the analysis bank, into a reconstructed signal $\hat{x}(n)$. An analysis filter bank followed by a synthesis filter bank constitutes what is known as a subband coding filter bank (SBC). A filter bank is called maximally decimated or critically sampled when all subbands are decimated by the same factor and the number of subbands is equal to the decimation factor. Figure 2.10 depicts an M-channel maximally decimated filter bank. Here, $X(z)$ represents the z-transform of the discrete-time sequence $x(n)$, defined as $X(z) = \sum_n x(n)z^{-n}$.

One class of two-channel filter banks is that of quadrature mirror filter (QMF) banks. In standard QMF, the amplitude response of the high-pass filter is a perfect mirror image of the response of the low-pass filter at a frequency of approximately $\omega = 2\pi/4$, and filters are power complementary [2].

Two-Channel PR Filter Banks

The two-channel filter bank [3–5,26] is the simplest type of filter bank, where the input signal is decomposed into one low-frequency subband and one high-frequency subband. In a two-channel maximally decimated filter bank (Figure 2.11), $H_0(z)$ and $H_1(z)$ are analysis low-pass and high-pass filters, respectively; and $F_0(z)$ and $F_1(z)$ are synthesis low-pass and high-pass filters, respectively. Outputs of analysis filters are downsampled by the factor of two, whereas sequences are upsampled by the factor of two before being fed into the synthesis filters.

In many applications, it is desirable to obtain a filter bank where subbands can be recombined to produce the output signal as closely as possible, in a certain well-defined sense, or even exactly the same as the original input signal. However, the recovered signal may not be an exact replica of the input signal and may suffer from amplitude or phase distortions and aliasing error because analysis filters $H_k(z)$, $k = 0, 1$ are not ideal. Figure 2.12 depicts typical frequency responses of nonideal analysis filters of a two-channel filter bank.

The filter bank is referred to as a PR filter bank, provided that the reconstructed signal is simply a delayed and scaled version of the input signal, that is, $\hat{x}(n) = ax(n – l)$, where a is any nonzero scaling constant and l is a positive integer. The reconstructed signal is completely free from aliasing and distortions.

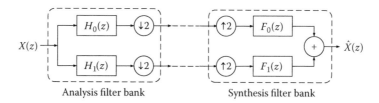

Analysis filter bank   Synthesis filter bank

FIGURE 2.11
Two-channel filter bank.

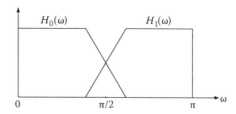

FIGURE 2.12
Frequency response of analysis filter bank.

A general one-dimensional two-channel PR filter bank is shown in Figure 2.11. The output $\hat{X}(z)$ of the filter bank can be expressed in terms of the input signal $X(z)$ and the filter transfer functions $H_k(z)$ and $F_k(z)$, as follows:

$$\hat{X}(z) = \frac{1}{2}\left\{H_0(z)F_0(z) + H_1(z)F_1(z)\right\}X(z)$$

$$+ \frac{1}{2}\left\{H_0(-z)F_0(z) + H_1(-z)F_1(z)\right\}X(-z)$$

$$= T_0(z)X(z) + T_1(z)X(-z) \tag{2.20}$$

The function $T_1(z)$ is loosely referred to as an *alias system function*. For alias cancellation, we require

$$T_1(z) = \frac{1}{2}\left\{H_0(-z)F_0(z) + H_1(-z)F_1(z)\right\} = 0 \tag{2.21}$$

If $T_1(z) = 0$, then the filter bank is a linear time invariant system, where the input is $X(z)$ and the output is $\hat{X}(z)$. In this case, the transfer function $T_0(z) = \dfrac{\hat{X}(z)}{X(z)}$ represents the distortions caused by the alias-free system. An alias-free filter bank does not suffer from any amplitude distortion if $T_0(z)$ is a (stable) all-pass function, whereas the filter bank is completely free from phase distortion if $T_0(z)$ is a linear phase finite impulse response (FIR) function.

If we want an alias-free system to be free from both amplitude and phase distortions, then $T_0(z)$ must be a monomial or simply a delay. Therefore, for distortion cancellation, we choose the distortion function as

$$T_0(z) = \frac{1}{2}\left\{H_0(z)F_0(z) + H_1(z)F_1(z)\right\} = C_0 z^{-l} \tag{2.22}$$

Thus, an alias-free filter bank, free from both amplitude and phase distortions, is a PR filter bank because the input and the output of the filter bank is related by $\hat{X}(z) = C_0 z^{-l}X(z)$, that is, output is simply a delayed and scaled version of the input.

Equations 2.21 and 2.22 can be expressed in a matrix equation as

$$\left[F_0(z)\ F_1(z)\right]\begin{bmatrix} H_0(z) & H_0(-z) \\ H_1(z) & H_1(-z) \end{bmatrix} = [2z^{-l}\ 0] \tag{2.23}$$

We may define two important matrices

$$\mathbf{H}^m(z) = \begin{bmatrix} H_0(z) & H_0(-z) \\ H_1(z) & H_1(-z) \end{bmatrix} \tag{2.24}$$

and

$$\mathbf{F}^m(z) = \begin{bmatrix} F_0(z) & F_1(z) \\ F_0(-z) & F_1(-z) \end{bmatrix} \tag{2.25}$$

which are referred to as *analysis modulation matrix* and *synthesis modulation matrix*, respectively.

For given analysis filters, $H_0(z)$ and $H_1(z)$, the unique solutions for synthesis filters, $F_0(z)$ and $F_1(z)$ exist, if the determinant of the analysis modulation matrix is nonzero, that is,

$$\det \mathbf{H}^m(z) \neq 0 \tag{2.26}$$

In this case, the unique solution for $F_0(z)$ and $F_1(z)$ can be obtained from Equations 2.23 and 2.24 as

$$F_0(z) = \frac{2z^{-l}}{\det \mathbf{H}^m(z)} H_1(-z) \tag{2.27}$$

$$F_1(z) = -\frac{2z^{-l}}{\det \mathbf{H}^m(z)} H_0(-z) \tag{2.28}$$

From Equations 2.27 and 2.28, it is clear that to obtain FIR solutions for synthesis filters, the determinant of the analysis modulation matrix must be a monomial (simple delay), that is,

$$\det \mathbf{H}^m(z) = cz^{-k}, c \in \mathbb{R}, k \in \mathbb{Z} \tag{2.29}$$

If $\det \mathbf{H}^m(z)$ is not a monomial, then synthesis filters $F_0(z)$ and $F_1(z)$ are, by default, infinite impulse response (IIR) because we cannot design filters $H_0(z)$ and $H_1(z)$ to completely cancel the resultant poles and to also meet frequency specifications. We shall deal with FIR filters throughout this chapter.

Product Filter and Design of PR Filter Banks

For the given analysis filters $H_0(z)$ and $H_1(z)$, the simplest and most straight forward choice of the synthesis filters $F_0(z)$ and $F_1(z)$, which satisfies the PR condition is:

$$F_0(z) = H_1(-z) \tag{2.30}$$

$$F_1(z) = -H_0(-z) \tag{2.31}$$

The question that is yet to be answered is, how do we chose analysis filters $H_0(z)$ and $H_1(z)$ in designing a PR filter bank? To answer the question, we define the *product filter* $P_0(z)$ as

$$P_0(z) \triangleq F_0(z)H_0(z) \tag{2.32}$$

$P_0(z)$ is a low-pass filter, its high-pass counterpart can be defined as

$$P_1(z) \triangleq F_1(z)H_1(z) \tag{2.33}$$

If synthesis filters are chosen as given by Equations 2.30 and 2.31 then

$$P_1(z) = -P_0(-z) \tag{2.34}$$

Substituting $P_0(z)$ and $P_1(z)$ from Equations 2.32 and 2.33, respectively, into Equation 2.22 simplifies the equation into:

$$2T_0(z) = H_0(z)\,F_0(z) + H_1(z)\,F_1(z) = P_0(z) - P_0(-z) = 2z^{-l} \tag{2.35}$$

Equation 2.35 infers that all the odd powers of $P_0(z)$ are zero except the term z^{-l}, and that l is an odd integer.

Thus, the design of two-channel PR filter banks boils down to three steps:

1. Design a low-pass product filter $P_0(z)$ that satisfies the conditions given by Equation 2.35
2. Factorize $P_0(z)$ into $F_0(z)$ and $H_0(z)$
3. Use Equations 2.30 and 2.31 to find $F_1(z)$ and $H_1(z)$

Note that there may be several different ways to factor the product filter $P_0(z)$ in $F_0(z)$ and $H_0(z)$. Hence, the factorization of $P_0(z)$ is not unique.

Let $p_0(n)$ be the impulse response corresponding to the product filter $P_0(z)$ and the sequence $p(n)$ is obtained by shifting $p_0(n)$ by l samples to the left-hand side to center it around the origin, that is,

$$p(n) = p_0(n + l) \tag{2.36}$$

Equivalently,

$$P(z) = z^l P_0(z) \tag{2.37}$$

Substituting Equation 2.37 into Equation 2.35 simplifies the PR condition to:

$$P(z) + P(-z) = 2 \tag{2.38}$$

Condition 2.38 implies that the product filter $P(z)$ is a half-band filter, in which all even powers of z in $P(z)$ are zero except for the constant term (coefficient of the term z^0).

Note that in most applications, $P(z)$ is a zero phase filter and the length of $P(z)$ is always $4m - 1$, $m \in \mathbb{Z}^+$ is a positive integer, that is, the sum of lengths of low-pass and high-pass filters of a FIR PR filter bank will always be $4m$. For more details, readers are referred to the work of Vetterli [5].

Half-Band Filter

A half-band filter is a zero phase FIR filter having impulse response $p(n)$ such that all even numbered samples of $p(n)$ are zero except for the sample at the origin.

$$p(2n) = \begin{cases} 1 & n = 0 \\ 0 & n \neq 0 \end{cases} \tag{2.39}$$

From the above definition, we obtain

$$P(z) + P(-z) = 2 \tag{2.40}$$

For real coefficient filters, substituting $z = e^{j\omega}$ in this equation we get:

$$P(e^{j\omega}) + P(e^{j(\pi+\omega)}) = 2 \tag{2.41}$$

Equation 2.41 suggests that the half-band filter is symmetric about frequency $\omega = \dfrac{\pi}{2}$, which justifies the name "half-band filter." Note that several methods such as the window method and the eigen filter approach are available in the literature to design the half-band filter.

Paraunitary Filter Banks

A square matrix $\mathbf{H}$, which satisfies

$$\mathbf{H}^H \mathbf{H} = \mathbf{H} \mathbf{H}^H = \mathbf{I}$$

is referred to as a *paraunitary matrix*, where $\mathbf{H}^H$ is the Hermitian transpose of the matrix $\mathbf{H}$. In other words, the inverse of matrix $\mathbf{H}$ is its Hermitian transpose $\mathbf{H}^H$. In case of real entries only, a paraunitary matrix is called an *orthogonal matrix*.

A matrix $\mathbf{H}(z)$ whose entries are polynomials in z is known as a *polynomial matrix*. A polynomial matrix $\mathbf{H}(z)$ is called a paraunitary matrix if it satisfies:

$$\mathbf{H}^H(z^{-1})\mathbf{H}(z) = \tilde{\mathbf{H}}(z)\mathbf{H}(z) = \mathbf{I} \ \forall z \tag{2.42}$$

Sometimes, a scale factor is also allowed on the left side of Equation 2.42. The matrix $\tilde{\mathbf{H}}(z) = \mathbf{H}^H(z^{-1})$ is said to be the paraconjugate of $\mathbf{H}(z)$. A simple example of this is the Haar filter bank, which is paraunitary.

In the two-channel filter bank (Figure 2.11) the filters are chosen as follows in an orthogonal filter bank:

$$F_0(z) = z^{-N} H_0(z^{-1})$$

$$F_1(z) = z^{-N} H_1(z^{-1})$$

where N is an odd integer, which is the order of the filters. In this case, modulation matrices $\mathbf{H}^m(z)$ and $\mathbf{F}^m(z)$ are paraunitary. The paraunitary property of modulation matrices leads to the following conditions on impulse response sequences corresponding to analysis low-pass and high-pass filters $h_0(n)$ and $h_1(n)$, respectively,

$$\sum_n h_0(n)h_0(n-2k) = \delta(k) \tag{2.43}$$

$$\sum_n h_1(n)h_1(n-2k) = \delta(k) \tag{2.44}$$

$$\sum_n h_0(n)h_1(n-2k) = 0 \tag{2.45}$$

These conditions are called *double shift orthogonality conditions*. The double shift orthogonality conditions also hold true for synthesis filters.

Thus, an orthogonal, FIR, real coefficient, PR filter bank has the following important features:

1. Filter lengths are the same and always even because double shift orthogonality conditions cannot be satisfied by odd length filters
2. Low-pass and high-pass channels of the analysis bank as well as the synthesis bank are power complementary. The set of filters $H_k(z)$,

$k = 0, 1$ is said to be a power complementary set if $|H_0(e^\omega)|^2 + |H_1(e^{j\omega})|^2 = c\ \forall\omega$, where c is a strictly positive constant

3. Analysis and synthesis modulation matrices are paraunitary
4. The synthesis bank is simply the transpose of the analysis bank. In other words, synthesis filters can be obtained by time reversal of the corresponding analysis filters

Example:

The simplest half-band filter (polynomial) has only three coefficients, that is, the polynomial has only two roots. The half-band polynomial of degree two is given as

$$P_0(z) = \frac{1}{2}z^{-1}(z + 2 + z^{-1}) \tag{2.46}$$

We can verify that it satisfies the condition for the half-band filter, that is,

$$P_0(z) - P_0(-z) = 2z^{-1} \tag{2.47}$$

In this particular case, there is only one way of factorization of the half-band filter $P_0(z) = z^{-1}H_0(z)\,H_0(z^{-1})$, which leads to:

$$H_0(z) = \frac{1}{\sqrt{2}}(1 + z^{-1}) \tag{2.48}$$

Choosing the filter $H_1(z)$ such that it is a conjugate quadrature filter (CQF) w.r.t. $H_0(z)$, that is,

$$H_1(z) = -z^{-1}H_0(-z^{-1}) \tag{2.49}$$

Results in

$$H_1(z) = \frac{1}{\sqrt{2}}(1 - z^{-1}) \tag{2.50}$$

We choose corresponding synthesis filters according to the following equations:

$$F_0(z) = H_1(-z) \tag{2.51}$$

$$F_1(z) = -H_0(-z) \tag{2.52}$$

This leads to the following set of synthesis filters:

$$F_0(z) = \frac{1}{\sqrt{2}}(1+z^{-1}) \tag{2.53}$$

$$F_1(z) = \frac{1}{\sqrt{2}}(-1+z^{-1}) \tag{2.54}$$

This choice of analysis and synthesis filters results in a PR filter bank. Cancellation of aliasing component is verified because

$$T_1(z) = \frac{1}{2}\left[F_0(z)H_0(-z) + F_1(z)H_1(-z)\right]$$

$$= \frac{1}{2}\left[H_1(-z)H_0(-z) - H_0(-z)H_1(-z)\right] = 0 \tag{2.55}$$

The distortion function simplifies to:

$$T_0(z) = \frac{1}{2}\left[H_0(z)F_0(-z) + H_1(z)F_1(-z)\right] = z^{-1} \tag{2.56}$$

Equation 2.56 confirms the absence of amplitude and phase distortions. Thus, the filter bank designed from the factorization of the half-band filter (Equation 2.46) is a PR filter bank.

We notice that this filter bank corresponds to the Haar MRA or the Haar wavelet system discussed previously (see Piecewise Constant Approximation and Haar Wavelet).

Linear Phase Biorthogonal Filter Banks

Two-channel orthogonal filter banks have many attractive attributes like power complementarity and identical analysis and synthesis bank (up to a time reversal). The linear phase property of the filters is one of the most desirable attributes in image compression applications. In case of two-channel, FIR PR filter banks having filter lengths of more than two, we cannot design orthogonal filter banks with a linear phase. However, in case of M-channel filter banks with $M > 2$, it is possible to design PR filter banks, which are orthogonal as well as linear phase.

To obtain two-channel linear phase filter banks, we ought to sacrifice orthogonality, which leads to the two-channel linear phase biorthogonal filter banks. Note, linear phase filter $H(z)$ of length N satisfies the following property: $H(z) = \pm z^{-(N-1)}H(-z^{-1})$.

In case of two-channel filter banks, if we choose $H_1(e^{j\omega}) = e^{-j\omega}F_0^*(e^{j(\omega+\pi)})$ and $F_1(e^{j\omega}) = e^{-j\omega}H_0^*(e^{j(\omega+\pi)})$, it leads to alias cancellation, and the distortion cancellation condition simplifies to:

$$H_0(e^{j\omega})F_0^*(e^{j\omega}) + H_0(e^{j(\omega+\pi)})F_0^*(e^{j(\omega+\pi)}) = 2 \tag{2.57}$$

where H^* denotes the complex conjugate filter of H.

In the time domain, biorthogonality conditions can be stated as

$$\sum_n h_l(n)f_m(n-2k) = \delta(k)\delta(l-m),\ l,m = 0,1, k \in \mathbb{Z} \tag{2.58}$$

In a two-channel linear phase biorthogonal filter bank, the length of all filters will be either even or odd. We can have the following two possible choices for analysis filters:

1. Both low-pass and high-pass filters are symmetric having odd lengths, the length difference is $4m + 2$, $m \in \mathbb{N}$
2. The low-pass filter is symmetric and the high-pass filter is antisymmetric, both having even lengths and their lengths differ by $4m$, $m \in \mathbb{N}$

For more details, readers are referred to the work of Vetterli and Herley [6] and Cohen et al. [7].

Maxflat Half-Band Filter

Here we illustrate the factorization of a special half-band filter. The filters obtained from the factorization yield orthogonal and biorthogonal wavelets with special properties. The close connection between the filters and corresponding wavelets will be explored in subsequent sections.

Consider the product filter $P_0(z)$:

$$P_0(z) = (1 + z^{-1})^{2p}R_{2p-2}(z) \tag{2.59}$$

Polynomial $R(z)$ of degree $2p - 2$ is chosen such that $P_0(z)$ satisfies the half-band filter property. In this case, $R(z)$ will be unique because the length of $R(z)$ and the number of the conditions on coefficients of $P_0(z)$ is equal to $2p - 1$. All odd powers of $P_0(z)$ are zero, except the center term with a coefficient value 1. Because $P_0(z)$ is symmetric, $p - 1$ terms are redundant and it has only p degrees of freedom. The product filter $P_0(z)$ is a linear phase, odd length, half-band polynomial of degree $4p - 2$. Thus, it has only $2p + 1$ nonzero terms, and p degrees of freedom. Because $P_0(z)$ has a maximum $2p$ number of zeros at $z = -1$; it is called a *maxflat filter* [8], which means the frequency response is maximally flat at the frequency $\omega = \pi$. The polynomial $P_0(z)$ is also called a *Lagrange half-band polynomial* (LHBP). Splitting $P_0(z)$ into $H_0(z)$ and $F_0(z)$ can give either linear phase filters or orthogonal filters. It cannot yield both, except in the case of Haar ($p = 1$). Note the spectral factorization of a normalized maxflat half-band filter, that is, one which is centered at the origin, yields Daubechies' orthogonal filter banks provided $R(e^{j\omega}) \geq 0$, which in turn lead to compactly supported wavelets.

Example:
Consider the maxflat half-band product filter for $p = 2$:

$$P_0(z) = (1 + z^{-1})^4 R_2(z) \tag{2.60}$$

Because $P_0(z) - P_0(-z) = 2z^{-3}$, all odd-indexed coefficients in $P_0(z)$ are 0, except for the center coefficient $p_0(3) = 1$. Let $R_2(z)$ be of the form

$$R_2(z) = a + bz^{-1} + az^{-2} \tag{2.61}$$

Substituting Equation 2.61 in 2.60, we get

$$\begin{aligned} P_0(z) = a &+ (4a + b)z^{-1} + (7a + 4b)z^{-2} + (8a + 6b)z^{-3} + (7a + 4b)z^{-4} \\ &+ (4a + b)z^{-5} + az^{-6} \end{aligned} \tag{2.62}$$

To satisfy the condition on $P_0(z)$, equating the coefficients of all odd powers of z with 0, except that z^{-3} carries the coefficient 1, yields

$$4a + b = 0 \tag{2.63}$$

$$8a + 6b = 1 \tag{2.64}$$

On solving Equations 2.63 and 2.64, we get

$$a = -\frac{1}{16}, \ b = \frac{1}{4}$$

Which yields the following $R_2(z)$ and $P_0(z)$:

$$R_2(z) = -\frac{1}{16} + \frac{1}{4}z^{-1} - \frac{1}{16}z^{-2} \tag{2.65}$$

$$P_0(z) = \frac{1}{16}(-1 + 9z^{-2} + 16z^{-3} + 9z^{-4} - z^{-6}) \tag{2.66}$$

We multiply $P_0(z)$ by $z^l = z^3$ to center it around the constant term z^0. The centering operation gives the polynomial $P(z)$:

$$P(z) = \frac{1}{16}(-z^3 + 9z^1 + 16z^0 + 9z^{-1} - z^{-3})$$ (2.67)

Thus, $P(z)$ is a half-band filter because all its even power terms are zero, except for z^0, which carries a coefficient of 1. Hence, the half-band condition $P(z) + P(-z) = 2$ is verified.

Having obtained a valid product filter of the form $P_0(z) = F_0(z)H_0(z)$, the objective is to find two unknown polynomial (FIR filters) $H_0(z)$ and $F_0(z)$. In case of a paraunitary filter bank, $F_0(z) = z^{-l}H_0(z^{-1})$, and this problem boils down to the problem of finding the spectral factor $H_0(z)$. There are many ways of distributing zeros to $H_0(z)$ and $F_0(z)$. In this particular case, $P_0(z)$ has six roots. The pole-zero plot of $P_0(z)$ is shown in Figure 2.13. Two roots, which come from $R_2(z)$, are $z_1 = 2 - \sqrt{3}$ and $z_2 = \dfrac{1}{z_1} = 2 + \sqrt{3}$. The other four roots are $z_3 = z_4 = z_5 = z_6 = -1$. Each factor is assigned at least one root at $z = -1$. Thus, we can have choices for $H_0(z)$ or $F_0(z)$. One can also exchange the roles of $H_0(z)$ and $F_0(z)$. Table 2.1 provides a list for the choices of $H_0(z)$ and $F_0(z)$ for some commonly used filters. In the table, type p/q denotes the fact that the length of the filters $H_0(z)$ and $F_0(z)$ are p and q, respectively. The last row of the table corresponds to Daubechies' compactly supported orthogonal wavelet.

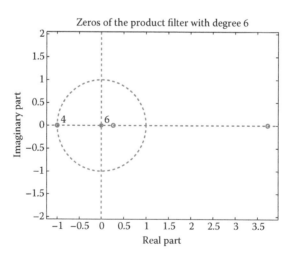

FIGURE 2.13
Pole-zero plot of sixth order maxflat half-band filter.

TABLE 2.1

Choice of $H_0(z)$ and $F_0(z)$

Type	Roots of $H_0(z)$	Roots of $F_0(z)$	Property
2/6	$(1 + z^{-1})$	$(1 + z^{-1})^3(z_1 - z^{-1})(z_2 - z^{-1})$	Linear phase
3/5	$(1 + z^{-1})^2$	$(1 + z^{-1})^2(z_1 - z^{-1})(z_2 - z^{-1})$	Linear phase
3/5	$(1 + z^{-1})(z_1 - z^{-1})$	$(1 + z^{-1})^3(z_2 - z^{-1})$	Linear phase
4/4	$(1 + z^{-1})^3$	$(1 + z^{-1})(z_1 - z^{-1})(z_2 - z^{-1})$	Linear phase
4/4	$(1 + z^{-1})^2(z_1 - z^{-1})$	$(1 + z^{-1})^2(z_1 - z^{-1})$	Orthogonal

Octave Filter Banks

The uniform filter bank consists of an array of filters which splits the input signal into subbands of equal bandwidth. The bandwidth of all filters is the same and center frequencies are uniformly spaced in frequency. The two-channel filter bank discussed so far decomposes the input signal into only two bands. In some applications, such as image compression, we require the portioning of the signal into more than two subbands of either equal or different bandwidths. We can split the spectrum into more than two bands, in general into M subbands. One can construct an M-channel filter bank by cascading two-channel filter banks appropriately. Iteration of two-channel filter banks at the output channels of preceding two-channel filter bank leads to a structure called a *tree structured filter bank* (Figure 2.14).

In various applications, it is essential to use filter banks that are nonuniform. In nonuniform filter banks, the given signal is split into subbands of different (unequal) bandwidths. Filter banks with exponentially spaced center frequencies and logarithmic bandwidths are of great importance in many applications. The well known example of this type of filter bank is the octave filter bank [6], which is closely related to dyadic wavelet filter banks. In an *octave filter bank* (Figure 2.14), the low-pass output of a two-channel filter

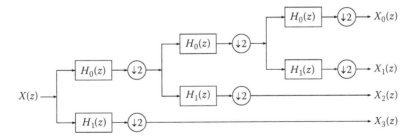

FIGURE 2.14
Tree structured four-channel filter bank.

bank is successively cascaded to a next stage two-channel filter bank. It is also called a *logarithmic filter bank* because on the logarithmic scale, channels have equal bandwidth. It is also called *constant-Q* or *constant relative band-width filter bank* because the quality factor "Q," which is defined as the ratio of the center frequency to the bandwidth, at each channel is the same. The name *octave filter bank* suggests a halving of frequency at every successive high-pass output. We will see in subsequent sections that dyadic wavelets are closely related to octave filter banks.

CWT versus DWT

In this section, we describe, in brief, the continuous wavelet transform (CWT) and the discrete wavelet transform (DWT).

Wavelets are basis function, suited to MRA of signals. We can use shifted and dilated versions of a finite energy prototype function $\psi(t)$ to obtain constant relative bandwidth analysis. This is often known as the WT [6]. The prototype function $\psi(t)$ is called the *mother wavelet*. The function $\psi(t) \in L^2(\mathbb{R})$ is a band-pass signal with zero average value and satisfies a condition known as the *admissibility condition*:

$$C_\psi = \int_{-\infty}^{\infty} \frac{|\psi(\omega)|^2}{|\omega|} d\omega < \infty \qquad (2.68)$$

The CWT or simply WT of a signal $f \in L^2(\mathbb{R})$ at time b and scale a is defined as

$$W_f(a,b) = \frac{1}{\sqrt{a}} \int_{\mathbb{R}} f(t)\psi^* \left(\frac{t-b}{a} \right) dt \qquad (2.69)$$

where $a \in \mathbb{R}^+$ and $b \in \mathbb{R},$ * denotes complex conjugation. Thus, the WT is a function of two continuous variables. The WT can be interpreted as the inner product of the signals $f(t)$ and $\psi_{a,b}(t) = \frac{1}{\sqrt{a}} \psi \left(\frac{t-b}{a} \right)$:

$$W_f(a, b) = \langle f(t), \psi_{a,b}(t) \rangle \qquad (2.70)$$

that is, the WT measures the similarity between the signal $f(t)$ and shifted and dilated versions of the mother wavelet $\psi(t)$. Note the factor $1/\sqrt{a}$ is used to normalize the $\psi_{a,b}(t)$. Here $\|\psi_{a,b}(t)\| = 1$, that is, the energy of $\psi_{a,b}(t)$ is unity.

The region of support of $W_f(a,b)$ is the region in the a–b plane in which the value of the WT is nonzero. In principle, the entire plane in $\mathbb{R}^+ \times \mathbb{R}$ can be the region of support. The WT is a redundant representation of the signal because it maps a one-dimensional signal into a two-dimensional one and the entire support of the WT need not be required to recover the signal $f(t)$. To remove redundancy, Yves Meyer discretized the parameters a and b, by taking

$$a = a_0^{-j}, b = a_0^{-j}k; \quad j,k \in \mathbb{Z}$$

This is equivalent to sampling of the WT at a discrete set of points in the a–b plane, which gives rise to a two-dimensional sequence $d_{j,k}$ commonly referred to as the DWT of the function $f(t)$. The CWT maps function $f(t)$ into a two-dimensional function $W_f(a,b)$ of continuous variables a and b, whereas DWT is a mapping of $f(t)$; $t \in \mathbb{R}$ into a two-dimensional sequence $d_{j,k}$. Thus, any signal $f(t) \in L_2(\mathbb{R})$ can be represented as an orthogonal wavelet series expansion if the sampling on the a–b plane is adequate:

$$f(t) = \sum_j \sum_k d_{j,k} a_0^{j/2} \psi(a_0^j t - k) \tag{2.71}$$

$$f(t) = \sum_j \sum_k d_{j,k} \Psi_{j,k}(t) \tag{2.72}$$

where $\psi_{j,k}(t) = a_0^{j/2}\psi(a_0^j t - k)$ and $d_{j,k} = \langle f(t), \psi_{j,k}(t)\rangle$. We call "$j$" the *scale factor* and "k" the *shift factor*.

The sampling of the WT corresponding to $a_0 = 2$ is called *dyadic sampling* because consecutive values of scales differ by a factor of 2. The resulting DWT is referred to as the *dyadic wavelet transform*. In particular, we represent a signal $f(t)$ as

$$f(t) = \sum_j \sum_k d_{j,k} 2^{j/2} \psi(2^j t - k) \tag{2.73}$$

Here, the redundancy of the CWT is removed by dyadic sampling of the CWT. We shall explain in subsequent sections, that the computation of wavelet coefficients is equivalent to a filtering operation followed by a downsampling operation. Further details can be found in the work of Burrus et al. [9].

Connection between Filter Banks and Wavelets

The concept of MRA blends the theory of wavelets and filter banks. Therefore, we shall summarize, in this section, the notion of MRA in brief. We also mention Mallat's algorithm to calculate wavelet coefficients efficiently. We shall also discuss the relationship between two-channel PR filter banks, wavelets, and scaling functions.

Multiresolution Analysis

The aim of MRA is to create a hierarchy of subspaces leading to the space of square integrable functions. Formally, a MRA is a sequence $\{V_j\}_{j\in\mathbb{Z}} = \{...V_{-2} \subset V_{-1} \subset V_0 \subset V_1 \subset V_2...\}$ of closed, nested subspaces that satisfy the following axioms: $\bigcap_{j\in\mathbb{Z}} V_j = \{0\}$ and $\bigcup_{j\in\mathbb{Z}} V_j$ leads to $L_2(\mathbb{R})$. In addition, V_j can be obtained from V_{j+1} by a contraction of factor two, that is, $f(t) \in V_j \Leftrightarrow f(2t) \in V_{j+1}; f(t) \in V_0 \Rightarrow f(t-k) \in V_0, \forall k \in \mathbb{Z}$; there exists a function $\phi(t) \in V_0$ such that $\{\phi\ (t-k)\}_{k\in\mathbb{Z}}$ constitutes an orthogonal basis for V_0.

Because $\phi(t) \in V_0 \subset V_1$ and the family of functions $\phi_{2,k}(t) = 2^{1/2}\phi(2t-k), k \in \mathbb{Z}$ span the space V_1, $\phi(t)$ can be represented in terms of the $\phi(2t-k)$ as

$$\phi(t) = \sum_k \sqrt{2}h_0(k)\phi(2t-k) \tag{2.74}$$

This is called the *dilation equation* for the dyadic scaling function. The coefficients $h_0(k)$ are called *scaling coefficients*.

The function $\phi(t)$ plays an important role in multiresolution decomposition and wavelet theory but it is not a wavelet. However, $\phi(t)$ is called a *scaling function* or a *father wavelet* by some authors. The orthogonal wavelet can be introduced as follows:

Because $V_j \subset V_{j+1}$, and V_{j+1} can be expressed by a direct sum as

$$V_{j+1} = V_j \oplus W_j, j \in \mathbb{Z} \tag{2.75}$$

The W_j is an orthogonal complement of the subspace V_j in the subspace V_{j+1}. The W_j represents additional information that is necessary to reach the approximation at the coarser scale represented by space V_j to the approximation at the finer scale represented by space V_{j+1}. The V_j's are called the *approximation subspaces* and W_j's the *detail subspaces*.

The space V_{j+2} can be decomposed into a direct sum $V_{j+2} = V_{j+1} \oplus W_{j+1}$. Thus, we can write

$$V_{j+2} = V_{j+1} \oplus W_{j+1} = V_j \oplus W_j \oplus W_{j+1}$$

By iterating the process, we can obtain the space of square integrable functions $L_2(\mathbb{R})$:

$$L_2(\mathbb{R}) = \underset{j \in \mathbb{Z}}{\oplus} W_j \qquad (2.76)$$

The central idea is to find one function $\psi(t)$ that can create all these spaces W_j. We now describe how this function, called the *wavelet* emerges. To do this, we need to identify a function $\psi(t) \in W_0$ such that family $\{\psi(t - k)\}_{k \in \mathbb{Z}}$ forms an orthogonal basis for W_0. Then, we can deduce that the families $\{\psi_{j,k}(t)\}_{j,k \in \mathbb{Z}} = \{2^{j/2}\psi(2^j t - k)\}_{j,k \in \mathbb{Z}}$ are orthogonal bases for the corresponding W_j, where W_j is the orthogonal complement of V_j in V_{j+1}. From Equation 2.76, we infer that $\{\psi_{j,k}(t)\}_{j,k \in \mathbb{Z}}$ forms an orthogonal basis for the space $L_2(\mathbb{R})$.

Because $\psi(t) \in W_0$ and $W_0 \subset V_1$, $\psi(t)$ can be represented as a linear combination of the functions $\phi(2t - k)$:

$$\psi(t) = \sum_k \sqrt{2} h_1(k) \phi(2t - k) \qquad (2.77)$$

This two-scale relation for $\psi(t)$ is the central equation of a dyadic MRA. This follows that any function $f \in L_2(\mathbb{R})$ can be written as

$$f(t) = \sum_j \sum_k d_{j,k} \psi_{j,k} \qquad (2.78)$$

where,

$$d_{j,k} = \langle f, \psi_{j,k} \rangle \qquad (2.79)$$

For more details, readers are referred to the work of Vetterli and Herley [6] and Burrus et al. [9].

Biorthogonal MRA

Until now, we defined only one sequence of nested subspaces $\{V_j\}_{j \in \mathbb{Z}}$ in $L_2(\mathbb{R})$ satisfying the axioms of MRA. We now define another family $\{\tilde{V}_j\}_{j \in \mathbb{Z}}$ of closed nested subspaces that also satisfy the axioms of MRA, where $\{\tilde{V}_j\}$ is the subspace spanned by a dual family of functions $\{\tilde{\phi}_{j,k}\}_{j,k \in \mathbb{Z}}$.

If

$$\int_{-\infty}^{\infty} \phi(t)\tilde{\phi}(t - k) dt = \delta(k), \quad k \in \mathbb{Z} \qquad (2.80)$$

then families $\{\psi_{j,k}(t)\}_{j,k\in\mathbb{Z}} = \{2^{j/2}\psi(2^j t - k)\}_{j,k\in\mathbb{Z}}$ and $\left\{\tilde{\psi}_{j,k}(t)\right\}_{j,k\in\mathbb{Z}} = \left\{2^{j/2}\tilde{\psi}(2^j t - k)\right\}_{j,k\in\mathbb{Z}}$ constitute two Riesz bases [7], with

$$\left\langle \psi_{j,k}, \tilde{\psi}_{\tilde{j},\tilde{k}} \right\rangle = \delta\left(j - \tilde{j}\right)\delta\left(k - \tilde{k}\right) \tag{2.81}$$

and we have

$$\tilde{W}_j \perp V_j$$

$$W_j \perp \tilde{V}_j$$

that is, subspace $\tilde{W}_j$ is orthogonal to the subspace V_j, and subspace W_j is orthogonal to the subspace $\tilde{V}_j$
where

$$W_j = \text{span}\{\psi_{j,k}\}_{j,k\in\mathbb{Z}} \text{ and } \tilde{W}_j = \text{span}\{\tilde{\psi}_{j,k}\}_{j,k\in\mathbb{Z}} \tag{2.82}$$

W_j and $\tilde{W}_j$ are complements of V_j and $\tilde{V}_j$ in V_{j+1} and $\tilde{V}_{j+1}$, respectively. Thus, we have four sets of spaces that form two hierarchies to span $L_2(\mathbb{R})$. Therefore, we can express any $f \in L_2(\mathbb{R})$ as

$$f = \sum_{j,k}\left\langle f, \psi_{j,k} \right\rangle \tilde{\psi}_{j,k} = \sum_{j,k}\left\langle f, \tilde{\psi}_{j,k} \right\rangle \psi_{j,k} \tag{2.83}$$

Associated with this biorthogonal MRA is a biorthogonal wavelet system consisting of two dual-scaling functions $\phi(t)$ and $\tilde{\phi}(t)$, and the corresponding wavelets $\psi(t)$ and $\tilde{\psi}(t)$ [6]. These four functions satisfy the following two-scale equations:

$$\phi(t) = \sqrt{2}\sum_k h_0(k)\phi(2t - k) \tag{2.84}$$

$$\tilde{\phi}(t) = \sqrt{2}\sum_k f_0(k)\tilde{\phi}(2t - k) \tag{2.85}$$

$$\psi(t) = \sqrt{2}\sum_k h_1(k)\phi(2t - k) \tag{2.86}$$

$$\tilde{\psi}(t) = \sqrt{2} \sum_k f_1(k)\tilde{\phi}(2t-k) \tag{2.87}$$

where $h_0(k)$, $f_0(k)$ and $h_1(k)$, $f_1(k)$ are finite length, real-valued sequences. We will call $h_0(k)$ and $f_0(k)$ the scaling sequences corresponding to primary scaling function $\phi(t)$ and dual scaling function $\tilde{\phi}(t)$, respectively. Similarly, sequences $h_1(k)$ and $f_1(k)$ are referred to as wavelet sequences corresponding to primary wavelet $\psi(t)$ and dual wavelet $\tilde{\psi}(t)$, respectively.

Equations 2.84, 2.85, 2.86, and 2.87 suggest a link between discrete-time sequences and continuous-time functions. They provide a prelude to the construction of continuous-time wavelet bases from discrete filters.

The discrete-time Fourier transform (DTFT) of these sequences satisfy the following conditions:

$$H_0(e^{j\pi}) = H_0(-1) = F_0(e^{j\pi}) = F_0(-1) = 0$$

$$H_0(e^{j0}) = H_0(1) = F_0(e^{j0}) = F_0(1) = \sqrt{2} \tag{2.88}$$

$$H_1(e^{j\pi}) = H_1(-1) = F_1(e^{j\pi}) = F_1(-1) = \sqrt{2}$$

$$H_1(e^{j0}) = H_1(1) = F_1(e^{j0}) = F_1(1) = 0 \tag{2.89}$$

$$\begin{bmatrix} F_0(e^{j\omega}) & F_0(e^{j(\omega+\pi)}) \\ F_1(e^{j\omega}) & F_1(e^{j(\omega+\pi)}) \end{bmatrix} \begin{bmatrix} H_0^*(e^{j\omega}) & H_1^*(e^{j\omega}) \\ H_0^*(e^{j(\omega+\pi)}) & H_1^*(e^{j(\omega+\pi)}) \end{bmatrix} = \begin{bmatrix} 2 & 0 \\ 0 & 2 \end{bmatrix} \tag{2.90}$$

Equation 2.90 is a frequency domain manifestation of Equation 2.58, presented in the context of biorthogonal filter banks. It again suggests a strong connection between discrete-time filters and continuous-time wavelets and scaling functions.

Thus, wavelets and PR filter banks are closely connected. We shall explain in the subsequent sections that the construction of compactly supported wavelets is equivalent to the design of PR (biorthogonal) filter banks possessing certain additional properties [6]. In this section, we note that this close connection leads to a computationally efficient algorithm to compute the DWT coefficients of a signal.

Fast Wavelet Transform

S. Mallat developed an algorithm called the fast wavelet transform (FWT) to compute wavelet coefficients. In this algorithm, only discrete sequences $h_0(n)$, $f_0(n)$ and $h_1(n)$, $f_1(n)$ are to be considered. These sequences correspond to filters of the underlying PR filter bank. According to the algorithm [10]:

We assume the input signal is a sequence, which corresponds to an approximation of a continuous-time signal $f(t)$, at a certain scale. Furthermore, we assume, without loss of generality, that this scale is $j = 0$. Thus, initially a function $f(t)$ is assumed to be in V_0, which can be expressed as $f(t) = \sum_k c_{0,k} \phi(t-k)$. We calculate the projections of $f(t)$ on the approximation spaces $\{V_j\}_{j \in Z}$ and on the spaces of details $\{W_j\}_{j \in Z}$ at lower resolutions, that is, at the scales corresponding to $(j < 0)$. Using the identity $V_0 = W_0 \oplus W_{-1} \oplus \dots \oplus W_{-j} \oplus V_{-j}$, we can synthesize $f(t)$ exactly from its coarser approximation at the scale $-j$ by adding the details. In particular, we can write

$$c_{j,k} = \langle f, \phi_{j,k} \rangle \tag{2.91}$$

and

$$d_{j,k} = \langle f, \psi_{j,k} \rangle \tag{2.92}$$

where, $\phi_{j,k}(t) = 2^{j/2}\phi(2^j t - k)$ and $\psi_{j,k}(t) = 2^{j/2}\psi(2^j t - k)$.

From Equations 2.74, 2.77, 2.91, and 2.92, we can deduce the following relations:

$$c_{j,k} = \sum_n c_{j+1,n} h_0(n-2k) \tag{2.93}$$

$$d_{j,k} = \sum_n d_{j+1,n} h_1(n-2k) \tag{2.94}$$

From Equations 2.93 and 2.94, we infer that the calculation of the coefficients proceeds hierarchically at each level, when we move from scale $j + 1$ to scale j, the sequence c_{j+1} is convolved with the sequence $h_0(-n)$ followed by downsampling by a factor of two. Similarly, to obtain the detailed sequence d_j at coarser level, the sequence c_{j+1} is convolved with the sequence $h_1(-n)$ followed by subsequent downsampling by a factor of two. Figure 2.15a depicts the analysis filter bank structure for the process of obtaining coefficients from higher to lower resolution. We can iterate this process on the computed approximation sequence and obtain new approximation and detail sequences at lower resolutions.

The higher resolution approximation sequence c_{j+1} can be computed from lower resolution approximation sequence c_j and detail sequence d_j using the relation:

$$c_{j+1,k} = \sum_n c_{j,n} h_0(k-2n) + \sum_n d_{j,n} h_1(k-2n) \tag{2.95}$$

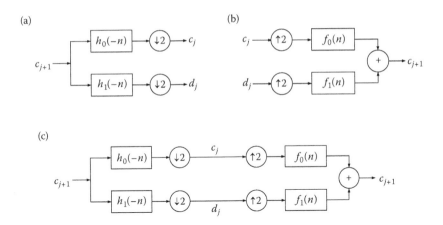

FIGURE 2.15
PR filter bank. (a) Analysis filter bank, (b) synthesis filter bank, and (c) Perfect reconstruction analysis–synthesis filter bank.

This relation can be realized by the synthesis filter bank shown in Figure 2.15b, which shows that to compute the sequence c_{j+1}, the sequences c_j and d_j are upsampled by a factor of two. The resulting sequences are convolved with the filters $h_0(n)$ and $h_1(n)$, respectively, and then added together. Iterations of this structure correspond to movement across a series of scales.

It is thus clear that the wavelet and the inverse WT system can be implemented in the form of analysis and synthesis filter banks, respectively (Figure 2.15c). This implies that there is a strong connection between the two-channel PR filter banks and the biorthogonal wavelet system or biorthogonal MRA. In fact, a dyadic wavelet decomposition of a discrete-time signal leads, naturally, to a filter PR bank approach to find approximation and detail coefficients at subsequent levels, if the approximating coefficients are given at particular level of the MRA. The two subjects of PR filter bank theory and WTs or MRA have evolved independently. It is to be noted that when two-channel PR filter banks are iterated for the construction of wavelets, they must satisfy some additional conditions beyond PR. Only then two-channel PR filter banks correspond to, what we call, the *valid wavelet filter banks*.

Wavelet Filter Banks: The Special Choice of PR Filter Banks

The dyadic DWT can always be implemented with two-channel PR filter banks. On the other hand, if we start with any PR filter bank and solve the two-scale dilation and wavelet equation, we will not necessarily obtain a solution that converges to smooth wavelets and scaling functions in $L_2(\mathbb{R})$. Every compactly supported orthogonal wavelet corresponds to a paraunitary filter bank. However, the converse is not true: a two-channel paraunitary filter bank may not necessarily correspond to an orthogonal wavelet.

Likewise, a biorthogonal WT is associated with a two-channel biorthogonal filter bank but the converse does not always hold true.

In other words, every two-channel PR filter banks cannot produce wavelets. It is associated with a valid wavelet filter bank provided some additional conditions are also satisfied by the underlying filters. A necessary condition that is to be satisfied by the low-pass filters of the two-channel PR filter bank is:

$$H_0(e^{j\pi}) = H_0(-1) = F_0(e^{j\pi}) = F_0(-1) = 0 \qquad (2.96)$$

that is, the low-pass filter must have at least one zero at the aliasing (Nyquist) frequency $\omega = \pi$. This property distinguishes wavelet filter banks from general PR filter banks. It is to be noted that the above property is necessary but not sufficient. In the sequel, we will state a necessary and sufficient condition, which ensures that the PR filter bank corresponds to a valid wavelet filter bank.

Generating Scaling Functions and Wavelets: The Cascade Algorithm

We will now discuss a method of generating scaling functions and wavelets from the set of filter coefficients $h_0(n)$ and $h_1(n)$ using successive approximation [6]. The two-scale dilation and wavelet equations are

$$\phi(t) = \sqrt{2}\sum_k h_0(k)\phi(2t-k) \text{ and } \psi(t) = \sqrt{2}\sum_k h_1(k)\phi(2t-k)$$

$\phi(t)$ and $\psi(t)$ are the solutions of the dilation and the wavelet equation, respectively. The two-scale equations can be used to find the approximate solution of the scaling function and wavelet, using the sequences $h_0(n)$ and $h_1(n)$. We can think of the two-scale equation as an iteration equation. The algorithm for constructing a wavelet or a scaling function is called *the cascade algorithm*. In this algorithm, the process starts with a initial approximation of scaling function $\phi^0(t)$, which is associated with the impulse response of the analysis low-pass filter $h_0(n)$, and the dilation equation is used recursively as given:

$$\phi^{i+1}(t) = \sqrt{2}\sum_k h_0(k)\phi^i(2t-k) \qquad (2.97)$$

If $\phi^i(t)$ converges to a limit $\phi(t)$, then the limit function solves the dilation equation, that is,

$$\phi(t) = \lim_{i \to \infty}\phi^i(t) \qquad (2.98)$$

To get the wavelet, one can use the equation

$$\psi(t) = \sqrt{2}\sum_{k} h_1(k)\phi(2t - k) \tag{2.99}$$

The cascade algorithm proceeds as follows:

First, we obtain the equivalent filters $H_0^i(z)$ and $H_1^i(z)$ after i steps of iteration (filtering and downsampling by two), using what are often called *noble identities* for upsampling and downsampling [6]:

$$H_0^{(i)}(z) = \prod_{p=0}^{i-1}\left(H_0(z^{2^p})\right). \tag{2.100}$$

$$H_1^{(1)}(z) = H_1(z) \tag{2.101}$$

$$H_1^{(i)}(z) = H_1(z^{2^{i-1}})\prod_{p=0}^{i-2}\left(H_0(z^{2^p})\right), i = 2, 3, \ldots \tag{2.102}$$

After i iterations, the equivalent filters involve upsampling by the factor 2^i. Note that the filter coefficients are normalized according to $\sum_{n} h_0(n) = 1$. The inverse z transforms of $H_0^{(i)}(z)$ and $H_1^{(i)}(z)$ are denoted by $h_0^{(i)}(n)$ and $h_1^{(i)}(n)$, respectively. Then, we associate the discrete-time iterated filter sequences $h_0^{(i)}(n)$ and $h_1^{(i)}(n)$ with the continuous time functions $\phi^{(i)}(t)$ and $\psi^{(i)}(t)$ as follows:

$$\phi^{(i)}(t) = 2^{i/2}h_0^{(i)}(n), \quad \frac{n}{2^i} \le t < \frac{n+1}{2^i} \tag{2.103}$$

$$\psi^{(i)}(t) = 2^{i/2}h_1^{(i)}(n), \quad \frac{n}{2^i} \le t < \frac{n+1}{2^i} \tag{2.104}$$

Here, the elementary interval is divided into intervals of size $1/2^i$. This rescaling is done to consider the fact that the length $L^{(i)} = (2^i - 1)(N - 1) + 1$ of the iterated filter $h_0^{(i)}(n)$ will become infinite as $i \to \infty$, where N is the length of the filter $h_0(n)$. Thus, the normalization ensures that the associated continuous-time function $\phi^{(i)}(t)$ remains compactly supported during the interval $[0, N - 1]$.

Thus, in a nutshell, to construct the scaling function from the sequence $h_0(n)$, the sequence is upsampled by a factor of two followed by convolution with the original sequence $h_0(n)$ and then rescaled. The resultant sequence is again upsampled by a factor of two followed by convolution with the sequence $h_0(n)$ and rescaling. Infinite iteration of this process will yield the scaling function.

Example:

Consider the case of the Haar filter bank where $h_0(n) = \{1, 1\}$ and $h_1(n) = \{1, -1\}$.

In this case, after three levels of iterations; we shall get following equivalent filters:

$$H_0^{(1)}(z) = H_0(z) = 1 + z^{-1}$$

$$H_1^{(1)}(z) = H_1(z) = 1 - z^{-1}$$

$$H_0^{(2)}(z) = H_0(z)H_0(z^2) = 1 + z^{-1} + z^{-2} + z^{-3}$$

$$H_1^{(2)}(z) = H_0(z)H_1(z^2) = 1 + z^{-1} - z^{-2} - z^{-3}$$

$$H_0^{(3)}(z) = H_0(z)H_0(z^2)H_0(z^4) = 1 + z^{-1} + z^{-2} + z^{-3} + z^{-4} + z^{-5} + z^{-6} + z^{-7}$$

$$H_1^{(3)}(z) = H_0(z)H_0(z^2)H_1(z^4) = 1 + z^{-1} + z^{-2} + z^{-3} - z^{-4} - z^{-5} - z^{-6} - z^{-7}$$

Note that filters $H_0^{(3)}(z)$ and $H_1^{(3)}(z)$ are preceded by upsampling by a factor of 8, filters $H_0^{(2)}(z)$ and $H_1^{(2)}(z)$ are preceded by upsampling by a factor of 4, and filters $H_0^{(1)}(z)$ and $H_1^{(1)}(z)$ are preceded by upsampling by a factor of 2.

Example:

Consider Daubechies' length 4 orthogonal filter bank, where the low-pass sequence is

$$h_0(n) = \{0.3415, 0.5915, 0.1585, -0.0915\}$$

and the corresponding high-pass sequence is

$$f_0(n) = \{-0.0915, -0.1585, 0.5915, -0.3415\}$$

Using the cascade algorithm, we can construct equivalent filters, scaling functions, and wavelets graphically. Figure 2.16a shows the original low-pass and high-pass sequences corresponding to equivalent

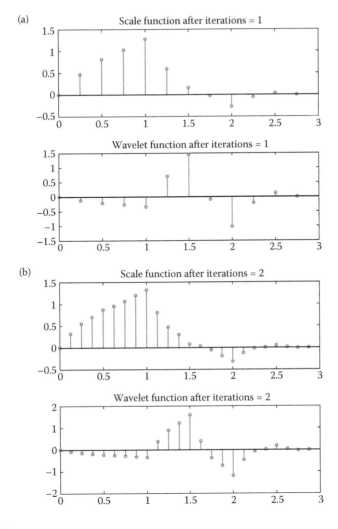

FIGURE 2.16
Daubechies' length 4 scaling function and wavelet using cascade algorithm.

filters $H_0^{(1)}(z)$ and $H_1^{(1)}(z)$, respectively, at iteration one. Figure 2.16b, c, and d depict equivalent filters $\{H_0^{(2)}(z)H_1^{(2)}(z)\}$, $\{H_0^{(3)}(z)H_1^{(3)}(z)\}$, and $\{H_0^{(4)}(z)H_1^{(4)}(z)\}$ corresponding to iteration-2, iteration-3 and iteration-4, respectively. Figure 2.16e plots approximate scaling function and wavelets functions obtained after eight levels of iterations.

A Necessary Condition for the Convergence in $L_2(\mathbb{R})$

We now need to discuss general conditions under which such an iteration would converge to smooth scaling functions and wavelets.

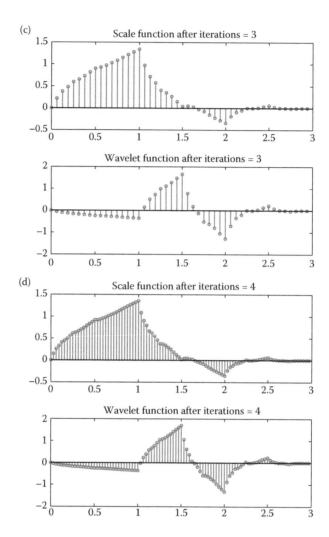

FIGURE 2.16 (Continued)
Daubechies' length 4 scaling function and wavelet using cascade algorithm.

Transition Matrix

An important matrix called the *transition matrix* is often used to determine the properties of wavelet systems [8]. First, we define a *Toeplitz matrix* $H_{ij} = h_{i-j}$ where h_k is the analysis low-pass filter whose coefficients are normalized as $\sum_k h_k = 1$. The matrix H is called the *convolution matrix* of filter h_k. Then, the transition matrix T is defined as [8]:

$$T = (\downarrow 2)\, 2HH^T \tag{2.105}$$

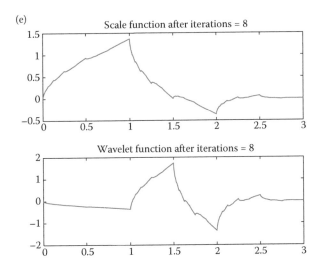

FIGURE 2.16 (Continued)
Daubechies' length 4 scaling function and wavelet using cascade algorithm.

where ($\downarrow$2) is the downsampling-by-2 operator. From Equation 2.105, the transition matrix T is obtained by removing alternate (even) rows of the symmetric positive definite matrix HH^T, which is the autocorrelation matrix of $h(n)$. For a real FIR filter, entries of HH^T are coefficients of the polynomial $|H(e^{j\omega})|^2$. The downsampling by a factor of 2 or equivalently removing even rows of the autocorrelation matrix HH^T, produces an aliasing term due to modulation by π. When this matrix T is operated on the sequence $f(n)$, the following output is produced in the frequency domain:

$$(TF)(\omega) = \left|H\left(\frac{\omega}{2}\right)\right|^2 F\left(\frac{\omega}{2}\right) + \left|H\left(\frac{\omega}{2}+\pi\right)\right|^2 F\left(\frac{\omega}{2}+\pi\right) \qquad (2.106)$$

The matrix T, also called the *Lawton matrix* was used by Lawton [11] to obtain necessary and sufficient conditions for an orthogonal wavelet basis. Eigenvalues of the matrix T also play an important role in determining regularity and smoothness properties of the associated scaling functions and wavelets.

We now review the necessary and sufficient conditions for the cascade algorithm to converge to functions in $L_2(\mathbb{R})$. For details, one can refer to the work of Strang and Nguyen [8].

The scaling function $\phi(t)$ *and the wavelet* $\psi(t)$ *converges in* $L_2(\mathbb{R})$ *if the absolute values of all eigenvalues of matrix T are less than 1 (except for the simple eigenvalue 1).*

For biorthogonal wavelet systems, we have to consider two transition matrices corresponding to the analysis filter bank as well as the synthesis filter bank. Because T is an infinite matrix, it is not possible to use this theorem for implementation as it is. However, an equivalent theorem of a similar nature has been developed by Strang and Nguyen [8] using the central sub-matrix of T of size $(2N - 1) \times (2N - 1)$, where N is the length of sequence $h(n)$.

Example:

Let us take an example of Daubechies' length 4 orthogonal filter bank, in which the low-pass filter is

$$h_0(n) = \{0.3415, 0.5915, 0.1585, -0.0915\}$$

The corresponding transfer matrix T is

$$T = \begin{bmatrix} -1/16 & 0 & 0 & 0 & 0 & 0 & 0 \\ 9/16 & 0 & -1/16 & 0 & 0 & 0 & 0 \\ 9/16 & 1 & 9/16 & 0 & -1/16 & 0 & 0 \\ -1/16 & 0 & 9/16 & 1 & 9/16 & 0 & -1/16 \\ 0 & 0 & -1/16 & 0 & 9/16 & 1 & 9/16 \\ 0 & 0 & 0 & 0 & -1/16 & 0 & 9/16 \\ 0 & 0 & 0 & 0 & 0 & 0 & -1/16 \end{bmatrix}$$

The eigenvalues of the transfer matrix are $\{1, 1/2, 1/4, 1/4, 1/8, -1/16, -1/16\}$. Thus, we notice that the absolute values of all eigenvalues are less than one except for one eigenvalue, which is unity. Therefore, the scaling function $\phi(t)$ and the wavelet $\psi(t)$ converge in $L_2(\mathbb{R})$, as shown in Figure 2.16e.

Example:

Consider the sequence $h_0(n) = \{1/2, 0, 0, 1/2\}$, which satisfies the joint conditions of double shift orthogonality as well as having a zero at $z = -1$. Thus, the sequence satisfies the necessary condition for convergence of cascade algorithm. The transition matrix associated with $h_0(n)$ is:

$$T = \begin{bmatrix} 1/2 & 0 & 0 & 0 & 0 & 0 & 0 \\ 0 & 0 & 1/2 & 0 & 0 & 0 & 0 \\ 0 & 1 & 0 & 0 & 1/2 & 0 & 0 \\ 1/2 & 0 & 0 & 1 & 0 & 0 & 1/2 \\ 0 & 0 & 1/2 & 0 & 0 & 1 & 0 \\ 0 & 0 & 0 & 0 & 1/2 & 0 & 0 \\ 0 & 0 & 0 & 0 & 0 & 0 & 1/2 \end{bmatrix}$$

The eigenvalues of the matrix T are $\{1, 1, 1/2, 1/2, 1/2, -1/2, -1\}$. In this case, more than one eigenvalue has unit magnitude. Thus, the necessary and sufficient condition for convergence of cascade algorithm fails. Therefore, the scaling function $\phi(t)$ and the wavelet $\psi(t)$ do not converge in $L_2(\mathbb{R})$ and an MRA cannot be constructed through these iterations.

Example:

Let us take two sequences $h_0(n)$ and $f_0(n)$:

$$h_0(n) = \{0.5375, 0.8625, -0.2875, -1.2250, -0.2875, 0.8625, 0.5375\}$$

$$f_0(n) = \{-2.8201, 4.5253, 0.1674, -4.2753, 5.8055, -4.2753, \\ 0.1674, 4.5253, -2.8201\}$$

One can easily verify that sequences satisfy condition of biorthogonality, that is, $\sum_n h_0(n) f_0(n - 2k) = \delta(k), k \in \mathbb{Z}$, and the z-transform of each sequence has two zeros at $z = -1$. The largest eigenvalues of transition matrices associated with $h_0(n)$ and $f_0(n)$ are 1.067 and 478.647, respectively. Therefore, the scaling function and the wavelets corresponding to the analysis as well as the synthesis filter banks do not converge in $L_2(\mathbb{R})$ as shown in Figure 2.17.

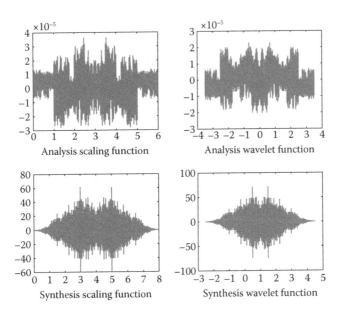

FIGURE 2.17
Irregular scaling function and wavelet.

Measure of Regularity

Regularity is the measure of smoothness of $\phi(t)$ and $\psi(t)$ in the sense of the degree of differentiability. There are two main notions used to characterize the smoothness of a function in the literature [12], that is, the *Holder regularity* (r_c) and the *Sobolev regularity* (s_c). The former is a pointwise measure whereas the latter is a global measure. r_c is harder to compute than s_c if the frequency response of the function is negative for some values of frequency. These two measures satisfy the inequality: $r_c \leq s_c \leq r_c + 1/2$. Hence, knowing s_c gives an idea about r_c as well.

The Sobolev regularity measure is defined as the smallest real number s_c such that $\hat{\phi}(\xi)$ satisfies

$$\int_{-\infty}^{\infty} \left|\hat{\phi}(\xi)\right|^2 (1+\xi^2)^s d\xi < \infty \tag{2.107}$$

for all $s < s_c$ [12].

To calculate this measure, we again use a result given by Strang and Nguyen [8]:

$$s_c = -\frac{\log \rho}{\log 4} \tag{2.108}$$

where $\rho = |\lambda_{max}(T)|$, is the absolute value of the maximum eigenvalue of T excluding the eigenvalues $\lambda = 1, \dfrac{1}{2}, \dfrac{1}{4}, \ldots, \left(\dfrac{1}{2}\right)^{2p-1}$. Here, p is the number of zeros of $H(z)$ at $z = -1$.

$$\rho = \max_{\lambda \neq 1, \frac{1}{2}, \frac{1}{4}, \ldots, \left(\frac{1}{2}\right)^{2p-1}} \left\{|\lambda(T)|\right\} \tag{2.109}$$

Example:

Consider Daubechies' length 8 orthogonal filter bank, in which the low-pass filter is

$h_0(n) = \{0.1629, 0.5055, 0.4461, -0.0198, -0.1323, 0.0218, 0.0233, -0.0075\}$

Note that the z-transform of the sequence has four zeros at $z = -1$. Eigenvalues of the transfer matrix associated with $h_0(n)$ are

$\{1, 0.5, 0.25, 0.125, 0.0853, 0.0625, 0.0312, 0.0163, 0.0156,$
$0.0152, 0.0078, -0.0024, -0.0024, -0.0459, -0.0583\}$

Thus, we notice,

$$\rho = \max_{\lambda \neq 1, \frac{1}{2}, \frac{1}{4}, \dots, \left(\frac{1}{2}\right)^{2p-1}} \left\{ \left| \lambda(T) \right| \right\} = 0.0853$$

Thus, the regularity of the scaling function or the wavelet corresponding to the filter is

$$s_c = -\frac{\log(0.0853)}{\log 4} = 1.7756$$

Design of Wavelet Filter Banks

One can use a direct approach on a continuous independent variable to
construct wavelets based on the axioms of MRA. There is another approach
based on filter banks in which discrete-time filters are iterated, and under
certain conditions, iterations lead to continuous wavelets. We can classify
wavelet filter bank design methods broadly into the following types:

1. Factorization of LHBP
2. Lattice structures
3. Lifting schemes
4. Parametrization of filter coefficients
5. Time–domain matrix–based methods

Using these methods, one tries to optimize several attributes pertaining
to wavelet filter banks depending on the application at hand. Commonly
used criteria are orthogonality, linear phase, number of vanishing moments,
regularity, frequency selectivity, energy compaction, coding gain, etc. [13].
Despite the requirement of good time–frequency localization, designing
bases (either functions or sequences) and filter banks that have good joint
time–frequency localization is still relatively unexplored. In the subsequent
section, we design wavelet filter banks via parametrization of filter coefficients, considering joint time–frequency localization of wavelets and associated filter banks as our optimality criterion.

The factorization of the half-band polynomial is one of the most popular
techniques used to design two-channel biorthogonal wavelet filter banks
(BWFBs). Cohen et al. designed compactly supported BWFBs by the factorization of the LHBP, which has the maximum number of zeros at $z = -1$ [7].

One of the limitations of this technique is that we do not have any freedom to control or optimize any attribute pertaining to the wavelet filter banks. Another limitation of this method is that the filters produced have irrational coefficients and therefore the corresponding DWT implementation requires infinite precision. This in turn increases the computational complexity. Another disadvantage of this method is that the number of spectral factors grows exponentially with respect to the degree of the LHBP and factorization of the LHBP becomes difficult when the number of zeros increases at $z = -1$.

Other popularly used methods to design two-channel linear phase, PR filter banks are discussed in the work of Vaidyanathan and Hoang [14] and Vetterli and Le Gall [15], in which filters are obtained directly from lattice parameters. In such structures, PR and linear phase properties are structurally imposed even when parameters are quantized. With these parameterizations, filter banks can be designed as an unconstrained optimization problem. Thus, the design problem is simply to find the lattice coefficients such that the objective function is optimized. The implementation of lattice structures is computationally efficient but the number of lattice coefficients increases linearly with the length of the filter involved. Therefore, the number of parameters to be optimized also increases correspondingly. One of the limitations of this method is the excessive dynamic range of the optimized coefficients. The filters are nonlinear functions of the lattice parameters. Consequently, the objective function is also nonlinear with respect to the lattice parameters. Therefore, we can design filters using numerical optimization techniques only.

There is another method to design two-channel wavelet filter banks called the lifting scheme [16]. In the lifting framework, starting with a small length filter, we can construct longer length filters very easily. The central idea is to build complicated biorthogonal systems by cascading simple and invertible stages. The advantage of the lifting scheme is that it requires fewer numerical operations than direct implementations. The limitation of this technique is that if rounding is performed on the parameters then, in general, almost all vanishing moments are lost.

Time–domain matrix–based design methods are suggested by Nayebi et al. [17], in which PR conditions are represented in terms of time–domain matrices and filter coefficients. This method uses optimization algorithms to yield good filters. The quadratic-constrained approach formulates the objective function and constraints in quadratic form. The advantage of this approach is that the derivatives and the Hessian of the cost function and constraints can be computed exactly. This is desirable in a minimization algorithm. One limitation of this method is that in cases where matrices associated with either objective functions or constraints are nonpositive definite, the optimization problem does not yield an exact or global solution. The other limitation of this method is that several desirable attributes (objective functions) pertaining to wavelet filter banks, such as time–frequency localization, regularity, and coding gain as well as constraints such as convergence of cascade algorithm in $L_2(\mathbb{R})$ cannot be expressed in quadratic form. Thus, in these cases, we cannot use this method directly.

To circumvent the above limitations of the LHBP, Tay [18] suggested a technique to design wavelet filter banks by reducing some of the zeros at $z = -1$. This gives some degree of freedom in designing the filter bank. However, Tay [18] constructed only 9/7 and 11/9 families of linear phase biorthogonal wavelet filers bank with rational and binary filter coefficients. In this chapter, we design higher-order wavelet filter banks using this technique. This enables us to have at least one free parameter, which is optimized to design linear phase biorthogonal discrete-time wavelet filter banks with better time–frequency resolution and regularity.

Design of Orthogonal Wavelets

The two-scale equation (Equation 2.74) shows that the coordinates of $\phi(t)$ with respect to the basis $\sqrt{2}\phi(2t - k)$ are the filter coefficients $h_0(n)$. Let, $h_0(n)$ be the impulse response of the unitary FIR filters satisfying the orthogonality conditions, that is,

$$\sum_n h_0(n)h_0(n - 2k) = \delta(k) \qquad (2.110)$$

and the system function $H_0(z)$ has p number of vanishing moments, that is, the filter $H_0(z)$ has the following form:

$$H_0(z) = \left(\frac{1 + z^{-1}}{2}\right)^p L(z) \qquad (2.111)$$

where $L(-1) \neq 0$, the length of the filter $H_0(z)$ is N, that is, $H_0(z)$ is an $N - 1$ degree polynomial. Because $H_0(z)$ has p number of zeros at $z = -1$, the degree of the polynomial $L(z)$ is $N - p - 1$. For $h_0(n)$ to be orthogonal to its even translates, it is required to satisfy $N/2$ orthogonality conditions, leaving at most $N/2$ degrees of freedom. Hence, the number of vanishing moments is limited by

$$1 \leq p \leq N/2 \qquad (2.112)$$

Daubechies [19] used degrees of freedom to design orthogonal filter banks and associated wavelets with a maximum number of vanishing moments. Patil et al. [20] have allowed a smaller number of vanishing moments and used the resulting extra degrees of freedom for other design purposes. Because these filters have a finite length impulse response of length N, the corresponding wavelets are compactly supported on the interval $[0, N - 1]$.

Once we have obtained a filter with the desired properties, we can derive the scaling and wavelet functions using the cascade algorithm, provided that the algorithm converges for this filter. The minimal conditions to be satisfied by the filter $h_0(n)$ are

- Double shift orthogonality:

$$\sum_n h_0(n)h_0(n - 2k) = \delta(k) \tag{2.113}$$

- At least one vanishing moment:

$$\sum_n n^k(-1)^n h_0(n) = 0 \text{ at least for } k = 0 \tag{2.114}$$

The above equations can be explicitly solved for the filter coefficients when the filter length is short.

Design of Compactly Supported Orthogonal Wavelets on the Interval [0, 1]

To design this orthogonal wavelet, which is compactly supported on the interval $[0, 1]$, the corresponding filter, $h_0(n)$, $n = 0, 1$, shall satisfy the following conditions:

$$h_0^2(0) + h_0^2(1) = 1 \tag{2.115}$$

and

$$h_0(0) = h_0(1) \tag{2.116}$$

These equations yield, $h_0(n) = \left\{ 1/\sqrt{2}, 1/\sqrt{2} \right\}$. Having obtained $h_0(n)$, the cascade algorithm may be used to construct wavelet and scaling functions. The sequence $h_0(n)$ leads to the Haar scaling function and the Haar wavelet function.

Design of Compactly Supported Orthogonal Wavelets on the Interval [0, 3]

To construct an orthogonal wavelet that is compactly supported over the interval $[0, 3]$, the corresponding filter $h_0(n)$ ought to have a length of four and at least one zero at $z = -1$. The filter must satisfy double shift orthogonality constraints to be orthogonal. If the filter has only one zero at $z = -1$, it will leave one degree of freedom. Let the filter have the following form:

$$H_0(z) = (1 + z^{-1})(1 + Az^{-1} + Bz^{-2})$$

$$H_0(z) = 1 + (1 + A)z^{-1} + (A + B)z^{-2} + Bz^{-3} \qquad (2.117)$$

Here, A and B are appropriate constants. The filter coefficients satisfy a double shift orthogonality condition so that the filter is orthogonal, leading to:

$$(A + B) + (1 + A)B = 0$$

which gives

$$A = \frac{-2B}{1+B} \qquad (2.118)$$

Using Equations 2.117 and 2.118, we can express the filter coefficient of the filter $H_0(z)$ in terms of free parameter B:

$$h_0(0) = 1, h_0(1) = \frac{1-B}{1+B}, h_0(2) = \frac{-B(1-B)}{1+B}, h_0(3) = B$$

It is to be noted that most of the B values will not lead to regular wavelets. If we choose $B = -2 - \sqrt{3}$, filter $H_0(z)$ will have two zeros at $z = -1$ and the remaining two zeros inside the unit circle of the z plane. This filter will lead to the Daubechies' wavelet emanating from a length 4 filter.

Design of Linear Phase Biorthogonal Filter Banks: Parametrization Technique

In the section Maxflat half-band filter, we illustrated that biorthogonal filter banks can be designed by factorizing a product filter, which satisfies the half-band condition. Most of the popular biorthogonal filter banks are designed by factorization of LHBP, which has a maximum number of zeros at $z = -1$. Cohen et al. [7] designed the class of biorthogonal filter banks, known as the CDF (Cohen–Daubechies–Feauveau) wavelet family, using this method. However, LHBP does not have any degree of freedom and thus, we do not control or optimize any attribute or criterion pertaining to the wavelet filter bank such as the time–frequency product (TFP). To obtain the degree of freedom, Tay [18], relaxes some of the vanishing moments of the LHBP and this freedom is used to obtain 9/7 and 9/11 wavelet filter banks with rational and binary coefficients, although this freedom could be used to optimize any other attributes. Liu et al. [21] presented the parametrization for linear phase biorthogonal filter banks with one free parameter. They used this freedom to obtain a filter bank with optimal coding gain. A parametric construction of linear phase biorthogonal filter banks with two free parameters is given by Liu et al. [22]. This freedom is used to design filter banks suitable for image coding. Based on the work of Tay [18], we design odd and even length filters of higher order with one, two, and three free parameters. We use this freedom to optimize the TFP of wavelets and the underlying filter bank.

Example: Design of Linear Phase Biorthogonal 7/5 Filter Bank with One Free Parameter

We start with a polynomial $P(z)$, which is a tenth degree symmetric polynomial with four zeros at $z = -1$. Because of this, the maximum available degrees of freedom is three $((10 - 4)/2)$, to obtain a biorthogonal PR filter bank, $P(z)$ must be a half-band polynomial, i.e., all its even powers $((z^4, z^{-4})$ and $(z^2, z^{-2}))$ should be zero. Therefore, out of three available degrees of freedom, two will be used to satisfy half-band conditions and one free parameter is still available. Thus, the coefficients of the polynomial can be expressed in terms of a free parameter. The polynomial $P(z)$ is then factored in such a way that each of its factors $H_0(z)$ and $F_0(z)$ has two zeros at $z = -1$, so that the necessary condition for the convergence of the cascade algorithm is satisfied.

Let the analysis low-pass and synthesis low-pass filters of the 7/5 linear phase biorthogonal filter bank be given by

$$H_0(z) = h_0 + h_1(z + z^{-1}) + h_2(z^2 + z^{-2}) + h_3(z^3 + z^{-3}) \tag{2.119}$$

$$F_0(z) = f_0 + f_1(z + z^{-1}) + f_2(z^2 + z^{-2}) \tag{2.120}$$

These polynomials in variable z are transformed into new polynomials in the variable Z, using the following change of variable:

$$Z = \frac{z + z^{-1}}{2}$$

The filter pair in the new variable Z can be expressed as

$$H_0(Z) = (Z + 1)(Z^2 + aZ + b) \tag{2.121}$$

$$F_0(Z) = (Z + 1)(Z + \alpha) \tag{2.122}$$

Here, a, b, and α are suitable constants. The product filter $P(Z) = F_0(Z)H_0(Z)$ is

$$P(Z) = Z^5 + \{2 + a + \alpha\}Z^4 + \{1 + 2a + b + (2 + a)\alpha\}Z^3 \tag{2.123}$$
$$+ \{a + 2b + (1 + 2a + b)\alpha\}Z^2 + \{b + (a + 2b)\alpha\}Z + b\alpha$$

The polynomial $P(Z)$ will be a valid half-band filter provided all its coefficients of even power are zero, that is,

$$2 + a + \alpha = 0 \tag{2.124}$$

$$a + 2b + (1 + 2a + b)\alpha = 0 \tag{2.125}$$

From Equations 2.124 and 2.125, a and b can be expressed in terms of parameter α, as follows:

$$a = -2 - \alpha \tag{2.126}$$

$$b = \frac{2 + 4\alpha + 2\alpha^2}{2 + \alpha} \tag{2.127}$$

The parameter α can be treated as the free parameter and all filter coefficients can be expressed in terms of the free parameter. Thus, we obtain

$$h_0 = \frac{1}{16}\left(\frac{2+\alpha}{\alpha(1+\alpha)}\right)$$

$$h_1 = -\frac{1}{8}\left(\frac{2+\alpha}{\alpha}\right)$$

$$h_3 = \frac{1}{16}\left(\frac{-2+3\alpha+4\alpha^2}{\alpha(1+\alpha)}\right)$$

$$h_4 = \frac{1}{4}\left(\frac{3\alpha+2}{\alpha}\right)$$

$$f_0 = \frac{1}{4}\left(\frac{3\alpha+2}{\alpha}\right)$$

$$f_1 = \frac{1}{2}(\alpha+1)$$

$$f_3 = \frac{1}{4}$$

To normalize coefficients of filters $H_0(z)$ and $F_0(z)$, all coefficients can be multiplied by a nonzero constant k. There is no restriction on the value of the free parameter α except for

$$\alpha \neq 0; \text{ and } \alpha \neq -1$$

The analysis and synthesis high-pass filters $H_1(z)$ and $F_1(z)$ can be obtained by quadrature mirroring the respective analysis and synthesis low-pass filters $H_0(z)$ and $F_0(z)$.

In this parametrization, we obtain rational coefficient filters for a rational value of the free parameter α. Attributes of the filters can be easily controlled by varying the value of the free parameter. In the case of higher order filters, we can relax more zeros in the LHBP and thus obtain more free parameters.

Time and Frequency Conflict: The Uncertainty Principle

The well known *Heisenberg uncertainty principle* in quantum mechanics states that both the position and the momentum of a free particle cannot be obtained simultaneously with arbitrarily fine resolution. A signal-processing interpretation of Heisenberg's uncertainty principle, which is known as the *time–frequency uncertainty principle* [23,24], essentially states that "a continuous-time signal and its Fourier transform cannot be simultaneously localized."

The more precisely we locate a signal in the time domain, the less precisely we can locate it in the frequency domain and vice versa. It is a fundamental limitation on how finely we can resolve any signal simultaneously in time and frequency.

Uncertainty Principle: Continuous-Time Signal

Let $f(x)$ be a normalized function in $L_2(\mathbb{R})$: $\int_{-\infty}^{\infty} |f(x)|^2 dx = 1$. Its Fourier transform is defined as $F(\xi) = \int_{-\infty}^{\infty} f(x)e^{-j\xi x}dx$. Here, $|f(x)|^2$ can be treated as a continuous probability density function because it follows: $|f(x)|^2 \geq 0$ and $\int_{-\infty}^{\infty} |f(x)|^2 dx = 1$. Hence, we define the first moment, which is called the *time center*, and the second moment, which is called the *time variance* [25].

The time variance or time spread of the function is defined as

$$\sigma_x^2 \triangleq \int_{-\infty}^{\infty} (x - x_0)^2 |f(x)|^2 dx \qquad (2.128)$$

where the time center x_0 is given as

$$x_0 \triangleq \int_{-\infty}^{\infty} x |f(x)|^2 dx \qquad (2.129)$$

Without loss of generality, we can assume that the function is centered around the origin in time, that is, $x_0 = 0$. This is because shifting in time does not affect the time variance.

We can define the *frequency variance* or *frequency spread* as

$$\sigma_\xi^2 \triangleq \frac{1}{2\pi} \int_{-\infty}^{\infty} (\xi - \xi_0)^2 |F(\xi)|^2 \, d\xi \tag{2.130}$$

where the frequency center ξ_0 is defined as

$$\xi_0 \triangleq \frac{1}{2\pi} \int_{-\infty}^{\infty} \xi |F(\xi)|^2 \, d\xi \tag{2.131}$$

It is important to modify the definition of frequency spread if the function is band-pass. Let $f(x)$ be a band-pass function, of which a wavelet is an example. We define the frequency spread for such a band-pass function as follows [25]:

$$\sigma_{\xi BP}^2 \triangleq \frac{1}{\pi} \int_{0}^{\infty} (\xi - \xi_0)^2 |F(\xi)|^2 \, d\xi \tag{2.132}$$

where the frequency center ξ_0 is defined as

$$\xi_{0BP} \triangleq \frac{1}{\pi} \int_{0}^{\infty} \xi |F(\xi)|^2 \, d\xi \tag{2.133}$$

Note that the limits of integration in the expressions for frequency spread and center are 0 and ∞ rather than $-\infty$ and ∞; and the multiplying factor is $1/\pi$ instead of $1/2\pi$. By considering only the positive frequency axis, we calculate the frequency localization of the function around its "true" center rather than 0, as would be obtained otherwise. For functions with a low-pass spectrum, the limits and the multiplying factor should be taken as $(-\infty, \infty)$ and $1/2\pi$ respectively.

For both cases, the uncertainty principle states that if $f(x)$ is a function such that $\sqrt{|x|} f(x) \to 0$ as $|x| \to \infty$, the product of the time spread and the frequency spread, known as the TFP, cannot be less than 0.25.

$$\Delta_f \triangleq \sigma_x^2 \sigma_\xi^2 \geq \frac{1}{4} \tag{2.134}$$

It implies that the TFP of wavelets and scaling functions is also lower bounded to 0.25. For discontinuous functions, the TFP $\to \infty$. For example, the TFP of the Haar wavelet is infinite. For more details, readers are referred to the work of Haddad et al. [25].

In the subsequent section, we design biorthogonal linear phase wavelets having optimal TFP as defined in this section. Because an exact analytic expression of the TFP of wavelets is not known in terms of filter coefficients, we have used numerical methods to arrive at the results.

Uncertainty Principle: Discrete-Time Signal

Many discrete-time measures have been proposed attempting to define time spread and frequency spread of a discrete signal or sequence, not necessarily in the variance sense [25,27–30]. In this section, we shall describe a time–frequency measure for a discrete-time sequence in the variance sense, analogous to continuous time signals as explained by Haddad et al. [25].

A Direct Extension from the Continuous-Time Measure

The uncertainty measure outlined by Haddad et al. [25] for discrete signals resembles the continuous time measure. We briefly review the measure here. If $h(n)$ is a sequence in l^2 ($\mathbb{Z}$) normalized as $\sum_{n=-\infty}^{\infty} |h(n)|^2 = 1$, and $H\left(e^{j\omega}\right) = \sum_{n=-\infty}^{\infty} h(n)e^{-j\omega}$ is its DTFT. The time variance of the sequence is defined as

$$\sigma_n^2 \triangleq \sum_{n=-\infty}^{\infty} (n - n_0)^2 |h(n)|^2 \tag{2.135}$$

where

$$n_0 \triangleq \sum_{n=-\infty}^{\infty} n |h(n)|^2 \tag{2.136}$$

The frequency variance, if $h(n)$ is a low-pass signal, is defined as

$$\sigma_\omega^2 \triangleq \frac{1}{2\pi} \int_{-\pi}^{\pi} \omega |H(\omega)|^2 \, d\omega \tag{2.137}$$

where

$$\omega_0 \triangleq \frac{1}{2\pi} \int_{-\pi}^{\pi} (\omega - \omega_0)^2 |H(\omega)|^2 \, d\omega \tag{2.138}$$

The definition is modified for band-pass signals in a similar manner as that in Equations 2.132 and 2.133. For high-pass signals, the spectrum is shifted by π to calculate the variance. A similar analysis as the continuous time case leads us to the following:

$$\text{Low-pass } \sigma_n^2 \sigma_\omega^2 \geq \frac{\left(1 - |H(-1)|\right)^2}{4}$$

$$\text{Band-pass } \sigma_n^2 \sigma_\omega^2 \geq \frac{(1-\mu)^2}{4} \text{ where } \mu = \frac{\omega_0}{\pi} |H(1)|^2 + \left(1 - \frac{\omega_0}{\pi}\right) |H(-1)|^2.$$

The lower bound in this definition of the uncertainty principle can be reduced to zero, unless appropriate care is taken about the spectral values at $\omega = 0$ and $\omega = \pi$. Interestingly, if the sequences possess certain properties, the uncertainty principle holds true for this class of discrete signals as well, and the TFP is lower bounded by 0.25; similar to that of continuous functions. For more details, readers are referred to the work of Haddad et al. [25].

Note that the low-pass filters of wavelet filter banks always have a zero at $z = -1$. Because of this special property, the uncertainty principle holds true for the impulse response of such filters, that is, the TFP cannot be made arbitrarily small and its lower bound is 0.25, that is

$$\Delta_h \triangleq \sigma_n^2 \sigma_\omega^2 \geq \frac{1}{4}, \quad H\left(e^{j\pi}\right) = 0 \tag{2.139}$$

In the next section, we design biorthogonal filter banks having optimized TFP for the underlying low-pass sequences.

Filter Bank Design: Time–Frequency Localization

The analysis of any signal calls for a trade-off in time and frequency localization. If we assume that the function is normalized, the uncertainty principle states that the product of the time spread (defined in the variance sense) and frequency spread is lower bounded by 0.25. To calculate the time–frequency localization of continuous functions, a variance-based measure is given by Haddad et al. [25], which we use to calculate the TFP of wavelets. Several different measures, not necessarily in the variance sense, have been proposed attempting to localize a discrete signal in time and frequency jointly [25,27–30]. Here, we use the measure given by Haddad et al. [25], which is a direct extension of the continuous measure, to calculate the TFP of discrete filters of the underlying filter bank. The measure states that for a class of sequences, possessing certain properties, the uncertainty principle holds true and the TFP is lower bounded by 0.25.

Dorize and Villemoes [31], have designed time–frequency resolution optimized continuous time orthonormal wavelets by expressing the optimization criterion in terms of the corresponding discrete filter coefficients. The work of Xie and Morris [32], attempts to design time–frequency localized orthonormal wavelets by optimizing the lattice structure parameters of the corresponding paraunitary filter bank. However, linear phase and orthogonality cannot be achieved together except in the case of the Haar filter bank. Hence, biorthogonal wavelets with a linear phase, which are very popular in image coding applications, cannot be designed using this method. Moreover, the literature on time–frequency optimized BWFBs is very sparse.

We now demonstrate a technique for the construction of linear phase BWFBs optimized for their time–frequency localization. The basic methodology employed is as follows: we construct a regular BWFB that has some free parameters as suggested by Tay [18]. We then optimize the free parameters to obtain time–frequency localized filter banks. However, the main difference in our method and that of Tay [18] is that we design higher-order filter banks with one, two, and three free parameters, whereas Tay [18] designed only 9/7 and 11/9 filter banks with two parameters. The main aim of Tay's work [18] was to design filter banks with rational and binary coefficients, whereas our aim was to design filter banks with optimal time–frequency localization.

In our earlier work [33], we designed BWFBs with very good time–frequency localization. The following additional improvements over our earlier work [33] are also part of the next discussions: to design optimal time–frequency localized filter banks, we adopt two different approaches. In one approach, our cost function is the TFP of wavelets, obtained by iterations of the cascade algorithm. In another approach, the TFP of the low-pass filters of the underlying filter bank is taken as the cost function. However, in our earlier work [33], the cost function is only the TFP of the iterated wavelet. We design even length as well as odd length filter banks. In addition, we evaluate and compare the performances of the optimal 9/7 filter banks in still image compression.

Our Design Methodology

In this section we present the design methodology used by us to design optimal filter bank, we adopt following procedure:

1. By using the parametrization technique, we construct linear phase BWFBs having one, two, and three degrees of freedom. Then, using the cascade algorithm, continuous-time wavelet functions and scaling functions $\phi(x)$, $\tilde{\phi}(x)$, $\psi(x)$, and $\tilde{\psi}(x)$ are generated.

2. In one approach, we minimize the TFP of the wavelets $\psi(x)$ or $\tilde{\psi}(x)$ and in the second approach, we minimize the TFP of the low-pass filters $h_0(n)$ or $f_0(n)$. The 9/7 filter bank designed with the former

approach is named as A-9/7 whereas the filter bank designed using the later approach is named as B-9/7.

3. To obtain filter banks with the minimum TFP, we first evaluated the TFP of wavelets and low-pass filters for several values of free parameters. It was found that both $\psi(x)$ and $\tilde{\psi}(x)$ turned out to be regular, that is, the transfer matrices T and $\tilde{T}$ on both the analysis and synthesis sides satisfied the conditions for convergence in $L_2(\mathbb{R})$ only if free parameters took values on a specific region in the space of free parameters. We evaluated the TFP on a set of finely spaced values in this region. From this set, taking the value of free parameters corresponding to the minimum TFP as the initial condition, we performed optimization using the optimization toolbox of MATLAB® with the cost function being $\Delta\psi(a, b, c)$ or $\Delta_{\tilde{\psi}}(a,b,c)$ in one approach and $\Delta_{h_0}(a,b,c)$ or $\Delta_{f_0}(a,b,c)$ in the other, with the constraint being the convergence condition. Where a, b, c are free parameters.

4. Simulations are performed to evaluate and compare the performance of the time–frequency optimized biorthogonal 9/7 filter banks in still image compression. The following steps are carried out to perform the experiments: level 1 to level 5 DWT is applied to the source images to generate a wavelet decomposition of the image. Symmetric extension is used at the image edge during the transform. Coefficients are encoded with set partitioning in hierarchical trees (SPIHT)* proposed by Pearlman and Said [34,35]. Experiments were carried out on three standard 8-bpp grayscale images, that is, Lena, Barbara, and Fingerprint at a wide variety of bit rates. To test rate-distortion performance, the objective measure peak signal to noise ratio (PSNR) in decibels is used.

Results and Comparisons

In Tables 2.2 and 2.3, examples of biorthogonal filter banks designed by the above method having time–frequency optimized analysis and synthesis wavelets are given. In these tables, we present the TFP (denoted as Δ) of wavelets and low-pass filters of biorthogonal filter banks designed by the above method. Additionally, the Sobolev regularity indices $S(\phi)$ and $S(\tilde{\phi})$ for the analysis and synthesis sides, respectively, are evaluated to give a measure of the smoothness of the obtained wavelets. Tables 2.4 and 2.5 present examples for biorthogonal filter banks, wherein the optimality criterion is the TFP of analysis low-pass and synthesis low-pass filters, respectively. Discrete low-pass filter coefficients $h_0(n)$ and $f_0(n)$ for optimal filter banks A-9/7 and B-9/7 have been tabulated in Table 2.6. Only the left-half part of the coefficients is shown; the remaining coefficients can be deduced by symmetry.

---

* The MATLAB® implementation of SPIHT codec used here was downloaded from the web site: http://www.mathworks.com/matlabcentral/fileexchange/4808-spiht.

TABLE 2.2

A Few Design Examples: Optimized Analysis Wavelet

Filter Bank	Δ_ψ	$\Delta_{\tilde\psi}$	Δ_{h_0}	Δ_{f_0}	$S(\phi)$	$S(\tilde\phi)$
9/7	0.394	4.129e3	0.271	0.751	1.878	1.92e-5
6/10	0.358	3.796e3	0.254	1.093	1.982	1.81e-4
8/8	0.326	2.497e3	0.335	0.854	2.000	0.014
10/10	0.302	554.380	0.311	0.459	2.658	3.13e-5
8/12	0.330	5.076e3	0.307	1.326	3.104	7.95e-5
9/15	0.339	3.692e3	0.252	1.441	3.214	6.31e-5
10/14	0.336	6.295	0.334	0.485	2.522	0.106
8/16	0.326	4.844e3	0.273	2.200	3.339	2.96e-5
11/17*	**0.280**	2.972e3	0.280	2.371	4.363	6.48e-5
10/18	0.296	3.06e3	0.283	2.063	2.418	3.05e-5

Figure 2.18 depicts the magnitude response of analysis and synthesis filters, product filter response, and plots of scaling functions and wavelets corresponding to optimal filter bank A-9/7. Figure 2.19 shows the same information corresponding to optimal filter bank B-9/7.

Table 2.7 compares the TFP of wavelets and low-pass filters of optimal filter banks A-9/7 and B-9/7 with the CDF-9/7 and optimal filter bank designed by Sharma et al. [33]. From the table, it can be seen that considerable improvement is achieved in the TFP of synthesis wavelets of filter bank A-9/7 over the previous work done on a similar theme [33]. It is also clear from the table that the TFP corresponding to synthesis low-pass filters of the optimal filter bank B-9/7 is superior to that of the CDF-9/7 filter bank.

Tables 2.8 through 2.12 show PSNR values of three experiment images reconstructed from level 1 to level 5 DWT decomposition, at various different bit rates at each level, using three optimal filter banks, namely, A-9/7,

TABLE 2.3

A Few Design Examples: Synthesis Wavelet

Filter Bank	Δ_ψ	$\Delta_{\tilde\psi}$	Δ_{h_0}	Δ_{f_0}	$S(\phi)$	$S(\tilde\phi)$
A-9/7	36.196	0.332	0.521	0.348	0.798	1.878
6/10[a]	499.262	**0.278**	0.413	0.373	0.042	2.000
8/8	1.705e3	0.337	0.933	0.356	0.264	1.997
10/10	0.443	0.537	0.315	0.904	2.628	1.673
8/12	8.611	0.300	0.517	0.369	0.099	2.000
9/15	1.118	0.304	0.398	0.719	1.671	2.686
10/14	0.482	0.330	0.327	1.151	2.887	2.070
8/16	0.544	0.292	0.331	0.890	1.973	2.393
11/17	0.348	0.467	0.304	2.731	4.118	1.916
10/18	0.471	0.303	0.341	2.179	1.713	2.920

[a] Indicates optimal filter bank having minimum value of the TFP, and corresponding TFP value is highlighted in bold fonts.

TABLE 2.4

A Few Design Examples: Analysis Filter

Filter Bank	Δ_ψ	$\Delta_{\tilde\psi}$	Δ_{h_0}	Δ_{f_0}	$S(\phi)$	$S(\tilde\phi)$
9/7	0.461	3.451e3	0.253	0.742	1.774	9.220e-5
6/10	0.669	3.803e3	0.251	1.051	1.876	6.920e-4
8/8	0.338	1.394e3	0.296	0.682	2.000	0.015
10/10	0.317	621.978	0.282	0.662	2.854	5.43e-6
8/12	0.382	3.247e3	0.253	1.005	1.764	0.022
9/15	0.339	3.741e4	0.252	1.451	3.231	1.011e-4
10/14	0.349	43.940	0.274	0.924	3.011	7.3e-5
8/16[a]	0.345	3.975e3	**0.251**	1.821	2.650	6.085e-4
11/17	0.312	3.558e3	0.257	1.582	4.151	3.011e-5
10/18	0.320	3.806e3	0.252	1.927	2.974	2.508e-5

[a] Indicates optimal filter bank having minimum value of the TFP, and corresponding TFP value is highlighted in bold fonts.

TABLE 2.5

A Few Design Examples: Synthesis Filter

Filter Bank	Δ_ψ	$\Delta_{\tilde\psi}$	Δ_{h_0}	Δ_{f_0}	$S(\phi)$	$S(\tilde\phi)$
B-9/7	31.973	0.748	0.740	0.256	0.792	1.602
6/10	148.337	0.354	0.370	0.334	0.2886	2.000
8/8[a]	2.055e3	0.528	0.986	**0.251**	0.131	1.803
9/11	19.442	0.669	0.800	0.256	0.886	1.667
10/10	0.887	7.982	0.716	0.272	1.638	0.968
8/12	8.431	0.411	0.470	0.310	9.178e-6	2.000
9/15	106.540	0.371	1.031	0.268	0.609	2.834
10/14	2.330	0.493	0.900	0.288	1.374	1.890
8/16	13.116	0.462	0.570	0.290	1.0e-5	2.109
11/17	0.490	0.556	0.848	0.326	2.037	2.227
10/18	3.448	0.418	0.951	0.279	1.258e-5	2.724

[a] Indicates optimal filter bank having minimum value of TFP, and corresponding TFP value is highlighted in bold fonts.

TABLE 2.6

Filter Coefficients: A-9/7 and B-9/7

k	$h_0(k)$	$f_0(k)$	k	$h_0(k)$	$f_0(k)$
0	6.7924e-1	5.1403e-1	0	7.8398e-1	4.6322e-1
1	2.5354e-1	2.9379e-1	1	2.9959e-1	2.4031e-1
2	-1.1172e-1	-7.0145e-3	2	-1.6813e-1	1.8390e-2
3	-3.5397e-3	-4.3786e-2	3	-4.9593e-2	9.6914e-3
4	2.2096e-2		4	2.6135e-2	
	A-9/7			B-9/7	

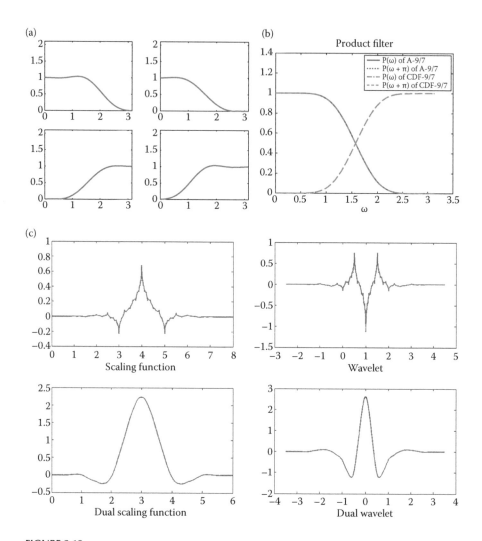

FIGURE 2.18
Plots for A-9/7. (a) Magnitude response of analysis and synthesis filters, (b) product filter magnitude response, and (c) scaling functions and wavelets.

B-9/7, and CDF-9/7. In terms of PSNR, no single BWFB performs best for all three images at each level. However, in most cases, optimal filter banks A-9/7 and B-9/7 (as a group) outperformed CDF-9/7.

For the image Fingerprint, CDF-9/7 always exhibits poor performance than either A-9/7 or B-9/7 (or both), at all levels. At levels 1 and 2, B-9/7 performs best whereas for levels 3 to 5, A-9/7 is the best.

For the images Lena and Barbara, A-9/7 performs best at all levels except at level 5, whereas B-9/7 performs worst at all levels. At level 5, CDF-9/7 performs marginally better than A-9/7. The maximum PSNR improvements

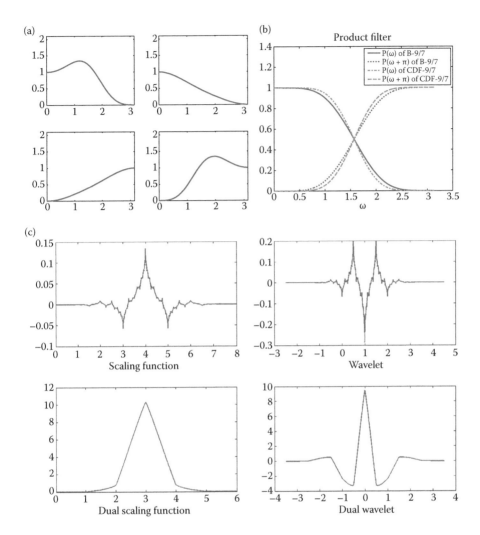

FIGURE 2.19

Plots for B-9/7. (a) Magnitude response of analysis and synthesis filters, (b) product filter magnitude response, and (c) scaling functions and wavelets.

TABLE 2.7

TFP and Sobolev Regularity of the Filter Banks

Filter Bank	Δ_ψ	$\Delta_{\tilde\psi}$	Δ_{h_0}	Δ_{f_0}	$S(\phi)$	$S(\tilde\phi)$
A-9/7	36.196	0.332	0.521	0.347	0.798	1.878
B-9/7	31.973	0.748	0.739	0.256	0.792	1.602
CDF-9/7	1.574	0.417	0.615	0.429	1.41	2.122
Best 9/7 in Sharma et al. [33]	3893.3	0.399	0.7423	0.2639	0.0002	1.874

TABLE 2.8

PSNR Results: Level 1 DWT Decomposition

Images	Filter Bank	Bit Rate					Average PSNR
		1	1/2	1/4	1/8	1/16	
Lena	A-9/7	11.50	9.76	7.40	6.37	6.03	8.21
	B-9/7	5.75	5.75	5.75	5.75	5.75	5.75
	CDF-9/7	11.30	9.64	7.37	6.35	6.02	8.14
Barbara	A-9/7	11.72	10.85	8.06	6.58	6.17	8.68
	B-9/7	5.89	5.89	5.89	5.89	5.89	5.89
	CDF-9/7	11.43	10.65	8.02	6.57	6.16	8.57
Fingerprint	A-9/7	4.58	4.58	4.58	4.58	4.58	4.58
	B-9/7	4.59	4.59	4.59	4.59	4.59	4.59
	CDF-9/7	4.58	4.58	4.58	4.58	4.58	4.58

TABLE 2.9

PSNR Results: Level 2 DWT Decomposition

Images	Filter Bank	Bit Rate					Average PSNR
		1	1/2	1/4	1/8	1/16	
Lena	A-9/7	22.07	12.74	11.35	9.86	7.39	12.62
	B-9/7	16.75	7.59	5.78	5.78	5.77	8.33
	CDF-9/7	21.55	12.37	11.01	9.48	7.32	12.25
Barbara	A-9/7	20.40	12.28	11.39	10.66	8.02	12.55
	B-9/7	16.06	8.26	5.92	5.92	5.90	8.41
	CDF-9/7	20.06	11.92	11.02	10.38	7.95	12.26
Fingerprint	A-9/7	15.62	6.21	4.58	4.58	4.58	7.11
	B-9/7	17.99	6.89	5.45	5.44	5.28	8.21
	CDF-9/7	14.66	6.09	4.58	4.58	4.58	6.89

TABLE 2.10

PSNR Results: Level 3 DWT Decomposition

Images	Filter Bank	Bit Rate					Average PSNR
		1	1/2	1/4	1/8	1/16	
Lena	A-9/7	35.84	27.63	20.98	12.47	11.18	21.62
	B-9/7	34.10	26.45	16.96	7.71	5.85	18.21
	CDF-9/7	35.78	27.27	20.51	11.98	10.72	21.52
Barbara	A-9/7	30.43	23.24	19.71	11.56	11.16	19.22
	B-9/7	28.62	22.67	16.34	8.24	5.92	16.35
	CDF-9/7	30.56	23.21	19.29	11.56	10.66	19.05
Fingerprint	A-9/7	29.65	23.15	16.86	11.19	10.53	18.28
	B-9/7	28.16	21.74	15.06	6.84	5.39	15.44
	CDF-9/7	29.48	22.77	16.52	10.56	9.95	17.86

TABLE 2.11

PSNR Results: Level 4 DWT Decomposition

Images	Filter Bank	Bit Rate					Average PSNR
		1	1/2	1/4	1/8	1/16	
Lena	A-9/7	39.04	35.12	30.34	24.35	19.70	29.71
	B-9/7	3835	33.97	29.30	23.62	17.10	28.47
	CDF-9/7	39.18	35.07	30.22	24.12	19.17	29.55
Barbara	A-9/7	34.51	28.79	24.29	21.67	18.54	25.56
	B-9/7	32.90	27.24	23.72	21.24	16.43	24.31
	CDF-9/7	34.70	29.02	24.40	21.53	18.12	25.55
Fingerprint	A-9/7	31.72	27.10	23.15	18.22	15.20	23.07
	B-9/7	29.90	25.97	21.96	16.96	13.83	21.72
	CDF-9/7	31.50	27.04	22.98	17.77	15.07	22.87

TABLE 2.12

PSNR Results: Level 5 DWT Decomposition

Images	Filter Bank	Bit Rate					Average PSNR
		1	1/2	1/4	1/8	1/16	
Lena	A-9/7	39.56	36.26	32.77	29.14	25.58	32.66
	B-9/7	38.74	35.50	31.66	28.60	25.17	31.93
	CDF-9/7	39.74	36.35	32.79	29.26	25.60	32.75
Barbara	A-9/7	35.30	30.03	26.02	23.53	22.15	27.40
	B-9/7	33.50	28.94	25.23	23.17	21.79	26.52
	CDF-9/7	35.53	30.33	26.40	23.79	22.20	27.65
Fingerprint	A-9/7	32.09	27.91	24.34	21.62	17.53	24.69
	B-9/7	30.25	26.78	23.30	20.19	17.37	23.59
	CDF-9/7	31.88	27.86	24.25	21.05	17.98	24.60

obtained using A-9/7 over CDF-9/7 are 0.37, 0.29, and 1.02 dB for Lena, Barbara, and Fingerprint, respectively. It is to be noted that PSNR improvement of 1.02 dB is very significant for the image Fingerprint. If we compare the performance of optimal filter banks designed by us, it is noticed that A-9/7 outperforms B-9/7 for images Lena and Barbara at all levels and all bit rates. It is to be noted that the Sobolev regularity measure of the synthesis sides of designed BWFBs is comparable to that of regularity of CDF-9/7 filter bank, whereas the Sobolev regularity of analysis sides is significantly better than the Haar filter bank. Figure 2.20 shows the rate distortion performance in terms of PSNR versus bit rate for all three images under evaluation, which are reconstructed from level 3 decomposition from A-9/7, B-9/7, and CDF-9/7 wavelet filter banks. From these figures, it is clear that the optimal filter bank A-9/7 outperforms CDF-9/7. Thus, we conclude that the filter banks

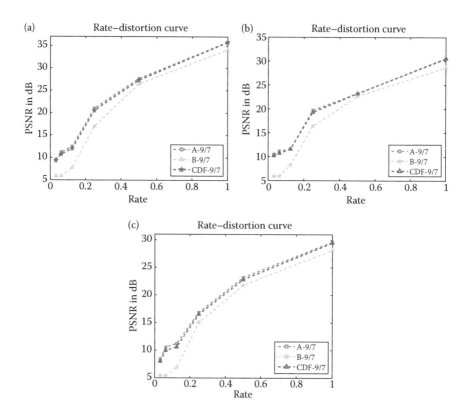

FIGURE 2.20
Rate distortion curves at level-3 decomposition. (a) Lena, (b) Barbara, and (c) Fingerprint.

designed by us are competitive with the CDF-9/7 filter bank in image compression, particularly for images with significant high-frequency details, such as the Fingerprint image.

Visual perceptual quality of the reconstructed images, coded by the three BWFBs, is also compared. To display various coding artifacts, we chose level 5 decomposition at a compression ratio of 32:1. Figures 2.21 through 2.23 depict the original images and the reconstructed images from three BWFBs under consideration. It is difficult to demonstrate the difference in perceptual quality of reconstructed images from optimal filter banks and the reference CDF-9/7 in print. However, at high compression ratios (>30:1), in a zoomed reproduction, it is observed that A-9/7 exhibits fewer ringing effects and the edges are better preserved than that in CDF-9/7 for all three images. For the sake of brevity, we do not go into more details here.

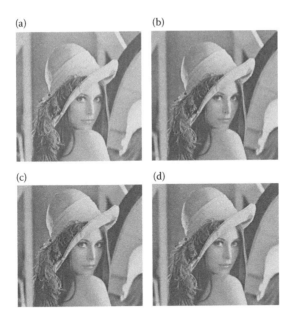

FIGURE 2.21
Lena images reconstructed from level 5 DWT decomposition at a compression ratio of 32:1. (a) Original, (b) using filter bank A-9/7, (c) using filter bank B-9/7, and (d) using filter bank CDF-9/7.

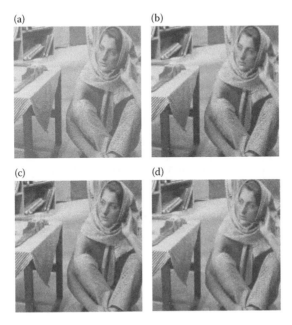

FIGURE 2.22
Barbara images reconstructed from level-5 DWT decomposition at a compression ratio of 32:1. (a) Original, (b) using filter bank A-9/7, (c) using filter bank B-9/7, and (d) using filter bank CDF-9/7.

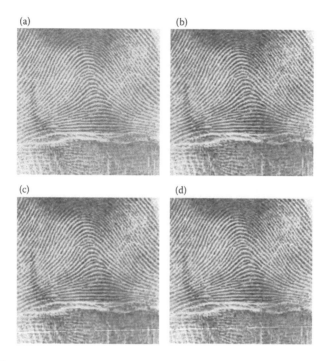

FIGURE 2.23
Fingerprint images reconstructed from level-5 DWT decomposition at a compression ratio of 32:1. (a) Original, (b) using filter bank A-9/7, (c) using filter bank B-9/7, and (d) using filter bank CDF-9/7.

Conclusions

In this chapter, we have reviewed the connection between filter banks and wavelets. Some of the popular design techniques to construct wavelet filter banks, with emphasis on parametrization methods to design biorthogonal filter banks, have been explained. We have designed linear phase BWFBs with optimized time–frequency localizations. These designs result in noticeable improvement over the previously designed filter banks in terms of the minimum value of the TFP of filters of the underlying filter banks and associated wavelets.

References

1. Gadre, V. M. Video lectures: Advanced digital signal processing—multirate and wavelets. Available: http://nptel.iitm.ac.in/courses/117101001 (Nov. 30, 2012).

2. Vaidyanathan, P. P. *Multirate Systems and Filter Banks*. Prentice-Hall: Englewood Cliffs, NJ, 1993.

3. Smith, M., and Barnwell, T., III. Exact reconstruction techniques for tree-structured subband coders. *IEEE Transactions on Acoustics, Speech and Signal Processing*, 34(3), 434–441, 1986.

4. Smith, M., and Barnwell, T., III. A new filter bank theory for time-frequency representation. *IEEE Transactions on Acoustics, Speech and Signal Processing* 35(3), 314–327, 1987.

5. Vetterli, M. A theory of multirate filter banks. *IEEE Transactions on Acoustics, Speech and Signal Processing* 35(3), 356–372, 1987.

6. Vetterli, M., and Herley, C. Wavelets and filter banks: Theory and design. *IEEE Transactions on Signal Processing* 40(9), 2207–2232, 1992.

7. Cohen, A., Daubechies, I., and Feauveau, J. C. Biorthogonal bases of compactly supported wavelets. *Communications on Pure and Applied Mathematics* 45(5), 485–560, 1992.

8. Strang, G., and Nguyen, T. *Wavelets and Filter Banks*. Wellesley-Cambridge Press: Wellesley, Massachusetts, 1996.

9. Burrus, C. S., Gopinath, R., and Guo, H. *Introduction to Wavelets and Wavelet Transforms—A Primer*. Prentice Hall: New Jersey, 1997.

10. Mallat, S. *A Wavelet Tour of Signal Processing, Second Edition (Wavelet Analysis & Its Applications)*. Academic Press: San Diego, California, 1999.

11. Lawton, W. Necessary and sufficient conditions for constructing orthonormal wavelet bases. *Journal of Mathematical Physics* 32, 57–61, 1991.

12. Unser, M., and Blu, T. Mathematical properties of the JPEG2000 wavelet filters. *IEEE Transactions on Image Processing* 12(9), 1080–1090, 2003.

13. Caglar, H., Liu, Y., Akanshu, A. N. Statistically optimized PR-QMF design. SPIE 1605 (Visual Communication and Image processing 91: Visual Communication) 86–94, 1991.

14. Vaidyanathan, P. P., and Hoang, P. Q. Lattice structures for optimal design and robust implementation of two-channel perfect-reconstruction QMF banks. *IEEE Transactions on Acoustics, Speech and Signal Processing* 36(1), 81–94, 1988.

15. Vetterli, M., and Le Gall, D. Perfect reconstruction FIR filter banks: Some properties and factorizations. *IEEE Transactions on Acoustics, Speech and Signal Processing* 37(7), 1057–1071, 1989.

16. Daubechies, I., and Sweldens, W. Factoring wavelet transforms into lifting steps. *Journal of Fourier Analysis and Applications* 4(3), 247–269, 1998.

17. Nayebi, K., Barnwell, T. P., and Smith, M. J. T. The time domain analysis and design of exactly reconstructing FIR analysis/synthesis filter banks. *ICASSP' 88, International Conference on Acoustics, Speech, and Signal Processing, 1988* 1735–1738, Apr 1990.

18. Tay, D. B. H. Rationalizing the coefficients of popular biorthogonal wavelet filters. *ISCAS '98, Proceedings of the 1998 IEEE International Symposium on Circuits and Systems*, May 31–Jun 3, 1998, 5, 146–149, 1998.

19. Daubechies, I. Orthonormal bases of compactly supported wavelets. *Communications on Pure and Applied Mathematics* XLI, 909–996, 1988.

20. Patil, B. D., Patwardhan, P. G., and Gadre, V. M. On the design of FIR wavelet filter banks using factorization of a half-band polynomial. *IEEE Signal Processing Letters* 15, 485–488, 2008.

21. Liu, Z., and Zheng, N. Parametrization construction of biorthogonal wavelet filter banks for image coding. *Signal Image Video Processing* 1(1), 63–76, 2007.

22. Liu, Z., and Gao, C. Construction of parametric biorthogonal wavelet filter banks with two parameters for image coding. *Signal Image Video Processing* 2(3), 195–206, 2008.

23. Gabor, D. Theory of communication. *Journal of the Institution of Electrical Engineers—Part III* 93, 429–457, 1946.

24. Papoulis, A. *Signal Analysis.* McGraw-Hill: New York, 1977.

25. Haddad, R. A., Akansu, A. N., and Benyassine, A. Time-frequency localization in transforms, subbands and wavelets: A critical review. *Optical Engineering* 32(7), 1411–1429, 1993.

26. Esteban, D., and Galand, C. Application of quadrature mirror filters to split band voice coding schemes. *ICASSP '77, IEEE International Conference on Acoustics, Speech, and Signal Processing,* May 1977, 2, 191–195, 1977.

27. Doroslovacki, M., Fan, H., and Djuric, P. M. Time-frequency localization for sequences. *Time-Frequency and Time-Scale Analysis, 1992. Proceedings of the IEEESP International Symposium,* Oct 4–6, 159–162, 1992.

28. Tay, P., Havlicek, J. P., and DeBrunner, V. A wavelet filter bank which minimizes a novel translation invariant discrete uncertainty measure. *Proceedings of Fifth IEEE Southwest Symposium on Image Analysis and Interpretation, 2002,* 173–177, 2002.

29. Przebinda, T., DeBrunner, V., and Ozaydin, M. Using a new uncertainty measure to determine optimal bases for signal representations. *ICASSP '99, IEEE International Conference on Acoustics, Speech, and Signal Processing,* Mar 15–19 1999, 1365–1368, 1999.

30. Donoho, D. L., and Stark, P. B. Uncertainty principles and signal recovery. *SIAM Journal on Applied Mathematics* 49(3), 906–931, 1989.

31. Dorize, C., and Villemoes, L. F. Optimizing time-frequency resolution of orthonormal wavelets. *ICASSP-91, International Conference on Acoustics, Speech, and Signal Processing,* Apr 14–17, 1991, 2029–2032, 1991.

32. Xie, H., and Morris, J. M. Design of orthonormal wavelets with better time–frequency resolution. *Proceedings of SPIE 2242, Wavelet Applications,* Mar 1994, 878–887.

33. Sharma, M., Kolte, R., Patwardhan, P., and Gadre, V. Time-frequency product optimized biorthogonal wavelet filter banks. *International Conference on Signal Processing and Communications (SPCOM),* Jul 18–21, 2010, 1–5, 2010.

34. Pearlman, W. A., and Said, A. Set partition coding: Part I of set partition coding and image wavelet coding systems. *Foundations and Trends in Signal Processing* 2(2), 95–180, Now Publishers: Delft, The Netherlands, 2008.

35. Pearlman, W. A., and Said, A. Set partition coding: Part I of set partition coding and image wavelet coding systems. *Foundations and Trends in Signal Processing* 2(3), 181–246, Now Publishers: Delft, The Netherlands, 2008.

3

Genesis of Wavelet Transform Types and Applications

N. Sundararajan and N. Vasudha

CONTENTS

Introduction 93
Short-Term Fourier Transform 94
Heisenberg's Uncertainty Principle... 95
Why Wavelets? 96
Wavelet Transform 97
Continuous Wavelet Transform 98
Discrete Wavelet Transform... 99
Wavelet Packet Transform 102
Fast Wavelet Transform... 103
Two-Dimensional Wavelet Transforms.. 104
Second-Generation Wavelets.. 107
Applications of WT to Potential Fields 110
References... 114

Introduction

The history of integral transforms began with D'Alembert in 1747. He described the oscillations of a violin string using a superposition of sine functions. Fourier proposed a similar idea for heat equations in 1807 and formulated the elegant mathematical tool—the Fourier transform (FT), which marked the beginning of the modern history of integral transforms. It serves as a benchmark for validating the existence of other well-known integral transforms such as the z-transform, the Hartley transform, the Walsh transform, the Laplace transform, the Hankel transform, the Mellin transform, the Hilbert transform, and the Radon transform. They all have a wide range of applications in a variety of fields in science and engineering and are invariably related to FT.

FT originally was used for the analysis of continuous signals and systems, which was later extended to discrete signals and systems. Computational aspects of the FT were further developed to speed up the computation, which led to fast Fourier transform (FFT; Cooley and Tukey 1965). Further advances in digital hardware technology, along with high-speed computational algorithms for the FT, resulted in its extensive application. Although the FT, with its wide range of applications, has its limitations because it may not be more suitable to processing nonstationary data in which it does not simultaneously ensure both frequency and time information—leading to the development of short-term Fourier transform (STFT).

Short-Term Fourier Transform

The FT gives the spectral content of the signal, devoid of any information regarding the time at which those spectral components appear. Such a result is appreciable only for stationary signals, whose frequency content does not change with time. Often, a particular spectral component occurring at a specific time instant can be of more interest as in the case of seismological data. This usually happens in the case of nonstationary signals, in which the frequency of the signal varies with time. To achieve time localization of frequency, a nonstationary signal is subjected through narrow windows, narrow enough that the portion of the signal seen from these windows is indeed stationary. This approach is called STFT, which is a modified version of FT. In STFT, the given signal is divided into small segments, in which the segments of the signal can be assumed to be stationary. For this purpose, a window function w is chosen, so that the width of the window is equal to the segment of the signal, in which its stationarity is valid. Thus, the STFT can be defined as (Gabor 1946; Allen and Rabinder 1977)

$$\text{STFT}(\tau, \omega) = \int_{-\infty}^{\infty} f(t)w^*(t - \tau)e^{-i\omega t}\, dt \qquad (3.1)$$

where $f(t)$ is the signal, $w(t)$ is the window function, τ is its width, and * indicates the complex conjugate. As can be seen from Equation 3.1, the STFT of the signal is nothing but the FT of the signal multiplied by a window function. For every τ and $f(t)$, a new STFT coefficient is computed. Thus, the STFT is an offshoot of the FT. In the same way, the short-time Hartley transform (STHT) can also be realized. In principle, the Hartley transform is a real replacement of the complex FT in which the amplitude is identical with the Fourier amplitude whereas the phase differs by 45 degrees (Bracewell 1983; Sundararajan 1995, 1997).

Heisenberg's Uncertainty Principle

STFT obeys "Heisenberg's uncertainty principle," which was originally applied to the momentum and location of moving particles, and can also be applied to the time–frequency information of a signal. According to this principle, one cannot know what spectral components exist at what instances of time, but one can know the time intervals in which a certain band of frequencies exist. This is known as time–frequency resolution. Hence, in STFT, if a window function equal to the length of the signal itself is used, we get the FT, which gives perfect frequency resolution but no time information. To obtain stationarity, the narrower the width of the window function is, the better the time resolution is; however, this results in poorer frequency resolution. Therefore, the choice of window function length becomes a trade-off between frequency resolution and time resolution. Resolution in time and frequency cannot be arbitrarily small because its product is lower bounded and hence results in time localization and frequency resolution that cannot be simultaneously determined to an arbitrary precision. This can be stated in terms of Heisenberg's uncertainty principle as

$$\Delta t \cdot \Delta f \geq 1 \tag{3.2}$$

where Δt and Δf are the uncertainties in time and frequency, respectively (Gabor 1946).

So, to analyze various types of behavior in a signal, the STFT would have to be taken several times, each time with a different size of window function. This is especially problematic when the location or duration (or both) of transient behavior are unknown, which is normally the case. Also, STFT suffers with the problem of frequency resolution. What is needed is a mapping that uses one initial window function that varies its size automatically so that all frequency behavior can be analyzed with just one pass (narrow window function at high frequencies and wide window function at low frequencies). This is where the wavelet transform (WT) comes in. The WT aims to address some of the shortcomings of the STFT. Instead of fixing the time and the frequency resolutions, Δt and Δf, one can let both resolutions vary in time–frequency plane to obtain a multiresolution analysis. This variation can be carried out without violating Heisenberg's inequality (given in Equation 3.2). In this case, the time resolution must increase as the frequency resolution decreases and the frequency resolution must increase as the time resolution decreases. This can be obtained by fixing the ratio of Δf over f to be equal to a constant c:

$$\frac{f}{\Delta f} = c \tag{3.3}$$

With this approach, the time resolution becomes arbitrarily good at high frequencies, whereas the frequency resolution becomes arbitrarily good at low frequencies. The WT allows the analysis of a signal that can locate energy in both time and frequency within the constraints of the uncertainty principle. Thus, the WT plays a significant role in signal analysis, overcoming the drawbacks encountered in FT/STFT (Hartley transform/STHT), when carried out with appropriate mother wavelets.

Why Wavelets?

As elucidated in the previous section, by replacing the window function of fixed width with a window function of variable width, we get the WT, a three-parameter transform with the new parameter for width of the window function, along with other two parameters (time and space). Hence, instead of decomposing the signal into sine and cosine functions of finite length, as in STFT, the signal is decomposed using wavelet functions of variable length. This allows the signal to be broken down into a collection of characteristic scales or frequency ranges that also preserve a sense of spatial location. The term "scale," in this context, describes the wavelength of a feature within the signal. Wavelet analysis has the advantage of being able to use larger windows to look at larger scale, lower-frequency parts of the signal, and vice versa. This ability to control the scale of observation has significant advantages with respect to the efficiency of data processing and also allows the processing to be varied with both scale and spatial location. Generally, the WT is a tool that divides data, functions, or operators into different frequency components and then studies each component with a resolution matched to its scale (Daubechies 1992). Therefore, the WT is expected to provide innovative, economic, and informative mathematical representations of many objects of interest (Abramovich et al. 2000).

Unlike the FT, the WT retains the locality of representation of the signal, which makes it possible to locally reconstruct the signal (Torrence and Compo 1998). Also, there is a possibility of reconstructing a part of the signal because there is a connection between the local behavior of the signal and the local behavior of its wavelet coefficients. Here, for obtaining a part of the reconstructed signal, it is necessary to consider the coefficients referring only to the corresponding subregion of the wavelet space. In case the wavelet coefficients involve random errors, they will act on the reconstructed signal only near the position of the perturbation, whereas the FT extends these errors to the entire signal being reconstructed. The FT is also especially sensitive to phase errors as a consequence of the alternating character of trigonometric series, whereas such a situation does not occur in the WT. Because wavelet coefficients contain combined information of the analyzing wavelet and the input signal, the selection of the analyzing wavelet determines what information should be extracted from the signal. Each wavelet has characteristic properties in time and frequency spaces. For the Fourier series, sinusoids are chosen as basis functions and then the properties of the resulting expansion are examined. For wavelet analysis, one poses the desired properties and then derives the resulting basis functions. Therefore, by using different wavelets, it is possible to additionally reveal and stress some or other properties of the signal that is being analyzed (Strang and Nguyen 1996).

Wavelet Transform

Although the WT owes its origin to the FT, the history of WT can be realized in a sequence of developments from Haar wavelets to Morlet wavelets. The Haar wavelet was proposed by the mathematician Alfred Haar in 1909 and is the first known literature regarding WT. The practical WT originated in 1980 with the geophysicist Jean Morlet, a French research scientist working on seismic data analysis (Morlet 1981, 1983; Goupillaud et al. 1984). Later, in 1981, the concept was proposed by Morlet followed by the physicist Alex Grossman, who invented the term *wavelet* in 1984. Following the first orthogonal wavelet, namely, the Haar wavelet, the mathematician Yves Meyer constructed the second orthogonal wavelet called the Meyer wavelet in 1985. In 1988, Stephane Mallat and Meyer proposed the concept of multiresolution analysis, which was followed by Ingrid Daubechies, who found a systematic method to construct the compact support orthogonal wavelet. In 1989, Mallat proposed the fast WT, which led to WT having numerous applications in the field of signal processing.

The WT can be defined as the correlation between the given signal and the mother wavelet $\psi(t)$ (such as Haar, Daubechies, Morlet, Mexican hat, etc.), which is obtained by translating and scaling the mother wavelet along the signal (Morlet 1983). Any function $\psi(t)$ that satisfies the conditions of zero mean (admissibility) and compact support (regularity) can be termed as a mother wavelet.

The translated and scaled mother wavelet is expressed as (Morlet 1983)

$$\psi_{\tau,s}(t) = \frac{1}{\sqrt{s}} \psi\left(\frac{t-\tau}{s}\right) \tag{3.4}$$

Here, s is a scale (wavelength) and τ is a translation parameter. The term translation is related to the location of the window and scale corresponds to a mathematical operation that either dilates or compresses $\psi(t)$. For larger values of s, the wavelet becomes more spread out and takes only the long time behavior of the signals, and when s is small, the wavelet becomes compressed and considers only the short-time behavior of the signal. Therefore, the WT provides a flexible timescale window that narrows when focusing on small-scale features and widens on large-scale features, analogous to a zoom lens. It is important to note that $\psi_{\tau,s}(t)$ has the same shape for all values of s. One may also interpret the WT as a mathematical microscope, in which the magnification is given by $1/s$ and the optics are given by the choice of wavelet $\psi(t)$ (Foufola-Georgoiu and Kumar 1994). There are quite a few types of WTs such as the continuous wavelet transform (CWT), discrete wavelet transform (DWT), stationary wavelet transform (SWT), and wavelet packet transform (WPT), which are briefly discussed hereunder.

Continuous Wavelet Transform

The CWT of a function $f(t)$ is given by Grossman and Morlet (1984) as

$$\mathrm{CWT}_f^{\psi}(\tau, s) = \int f(t)\psi_{\tau,s}^{*}(t)dt \qquad (3.5)$$

where $\psi_{\tau,s}^{*}(t)$ is the complex conjugate of the wavelet $\psi_{\tau,s}(t)$ (Daubechies 1992). The CWT is a correlation between a wavelet at different scales and the signal $f(t)$ (the data to be analyzed) with the scale (or the frequency) being used as a measure of similarity (Polikar 1996). It is computed by changing the scale of the analysis window function by shifting the window function in time, multiplying by the signal, and integrating overall times. Once a wavelet (generally referred to as the mother wavelet) is chosen, wavelet coefficients are computed for dilated and compressed versions of the wavelets.

The inverse CWT is given by (Burrus et al. 1998)

$$f(t) = \frac{1}{c_{\psi}} \int_{s} \int_{\tau} \mathrm{CWT}_f^{\psi}(\tau, s) \frac{1}{s^2} \psi\left(\frac{t-\tau}{s}\right) d\tau\, ds \qquad (3.6)$$

where c_{ψ} is constant, which depends on the choice of the mother wavelet.

The CWT is an energy-preserving transformation, that is,

$$\int_{-\infty}^{\infty} |f(t)|^2 \, dt = \frac{1}{c_{\psi}} \int_{-\infty}^{\infty} \int_{0}^{\infty} |\mathrm{CWT}_f^{\psi}(\tau, s)|^2 \, s^{-2} \, ds\, dt \qquad (3.7)$$

Qian and Chen (1996) expressed the function $\left|\text{CWT}_x^\psi(\tau,s)\right|^2$ as a scalogram. A scalogram provides an unfolding of the characteristics of a process in the scale–space plane.

Furthermore, there is a possibility of correcting the errors that are present in the wavelet coefficients of continuous transform, thanks to the redundancy inherent in CWT, which provides a way of using this method to investigate time series with gaps in the data. However, the redundancy in most of the information related to closed scales or times result in a high computational cost and therefore could be solved using discrete values of scale and translation. It is well known that Nyquist's sampling rate is the minimum sampling rate that allows the original continuous signal to be reconstructed from its discrete samples. Expressing the above discretization procedure in mathematical terms, the scale discretization can be expressed as $s = s_0^j$, and the translation discretization as $\tau = k \cdot s_0^j \cdot \tau_0$ where $s_0 > 1, \tau_0 > 0$.

Hence, the discrete wavelet is given as (Daubechies 1992)

$$\psi_{j,k}(t) = s_0^{-j/2} \psi\left(s_0^{-j}t - k\tau_0\right) \tag{3.8}$$

The WT obtained by using the discrete wavelet $\psi_{j,k}(t)$ is known as the discretized version of CWT and is expressed as

$$\text{DCWT}_f^\psi(j,k) = \int f(t)\psi_{j,k}^*(t)dt \tag{3.9}$$

Using the discrete wavelet $\psi_{j,k}(t)$ and the appropriate choices of s_0 and τ_0, we can also completely characterize a signal $f(t)$. These discrete wavelets, which provide complete representation of the signal $f(t)$, are called wavelet frames. The necessary and sufficient condition for this is that the wavelet coefficients $\text{DCWT}_f^\psi(j,k)$ satisfy

$$A\|f\|^2 \leq \sum_j \sum_k \left|\text{DCWT}_f^\psi(j,k)\right|^2 \leq B\|f\|^2 \tag{3.10}$$

where $\|f\|^2$ denotes the energy of the signal $f(t)$, and $A > 0$ and $B < \infty$ are constants characteristic of the wavelet and the choice of s_0 and τ_0, which can be determined numerically.

Discrete Wavelet Transform

Although the discretized version of CWT enables the computation of the CWT for discrete data and is the one used for practical applications, there exists another type of WT that is also called the DWT. As a matter of fact, the discretized version of CWT is simply a sampled version of the CWT, and the information it provides is highly redundant as far as the reconstruction of the signal is concerned. The DWT, on the other hand, provides sufficient information for analysis and synthesis of the original signal, with a significant reduction in computation time. The DWT is considerably easier to implement when compared with the CWT because it is based on the concept of multiresolution.

In the DWT, a timescale representation of a digital signal is obtained using digital filtering techniques. The signal is analyzed at different frequency bands with different resolutions by decomposing it into a coarse approximation and detailed information. The decomposition of the signal into different frequency bands is simply obtained by successive high-pass and low-pass filtering of the time domain signal. To begin with, the original signal is passed through a half-band, high-pass filter and a low-pass filter. After filtering, the signal is subsampled by 2, simply by discarding every other sample.

This decomposition halves the time resolution because only half the number of samples now characterizes the entire signal. However, this operation doubles the frequency resolution because the frequency band of the signal now spans only half the previous frequency band, effectively reducing the uncertainty in the frequency by half. The above procedure, which is also known as subband coding, can be repeated for further decomposition. The low-pass filter output is then filtered once again for further decomposition. At every level, the filtering and subsampling will result in half the number of samples (and hence half the time resolution) and half the frequency band (and hence double the frequency resolution). This process continues until only two samples are left.

The DWT employs two sets of functions, called scaling and wavelet functions, which are associated with low-pass and high-pass filters, respectively. With each mother wavelet $\psi(t)$, there is an associated scaling function, $\phi(t)$ such that the wavelet and the scaling functions are orthogonal and have the following relations (Burrus et al. 1998):

$$\phi(t) = \sqrt{2} \sum_k c_k \phi(2t - k) \tag{3.11}$$

$$\psi(t) = \sqrt{2} \sum_k (-1)^k c_{1-k} \phi(2t - k) \tag{3.12}$$

where c_k is a scaling function coefficient.

The detailed wavelet coefficients are denoted by $\mathrm{DWT}_f^\psi(j, k)$ and these coefficients represent the subsampled output of a high-pass filter, which can be alternatively calculated by translating and dilating the mother wavelet along the signal given by

$$\text{DWT}_f^\psi(j,k) = \int f(t)\psi_{j,k}(t)dt \tag{3.13}$$

where $\psi_{j,k}(t) = 2^{-j/2}\psi(2^{-j}t - k)$.

The scaling functions, which represent subsampled outputs of low-pass filters, can be used to calculate the approximation coefficients of the signal $f(t)$ in the same way as the detailed coefficients are calculated, that is, by replacing $\psi_{j,k}(t)$ with $\phi_{j,k}(t)$ in Equation 3.13.

The approximation coefficients contain information about the mean behavior of the signal. Thus, the detail coefficients and approximation coefficients together constitute the DWT. Furthermore, with the use of both detail and approximation coefficients, the signal can be reconstructed with the inverse DWT given by

$$f(t) = a_{j_0}(t) + \sum_{j=-\infty}^{j_0} d_j(t) \tag{3.14}$$

The details of the signal at level j also represent the upsampled and subsequently high-pass–filtered detail coefficients.

$$d_j(t) = \sum_{k=-\infty}^{\infty} \text{DWT}_f^\psi(j,k)\psi_{j,k}(t) \tag{3.15}$$

Furthermore, the approximation of the signal at level j_0, which is the result of the upsampled and subsequently low-pass–filtered approximation coefficients are given as

$$a_{j_0}(t) = \sum_{k=-\infty}^{\infty} \text{DWT}_f^\phi(j_0,k)\phi_{j_0,k}(t) \tag{3.16}$$

The relation between approximation and detail at one level and the approximation at the next level is expressed as

$$a_{k-1}(t) = a_k(t) + d_k(t) \tag{3.17}$$

The detail coefficients $\text{DWT}_f^\psi(j,k)$ contain the same information as the reconstructed details d_k, and the same is valid for the approximation coefficients $DWT_f^\phi(j,k)$ and the reconstructed approximation a_k. Hence, it is possible to use either the coefficients or the reconstructed approximation and details when analyzing the signal. The advantage with these coefficients (both approximate and detail) is that their number is decreasing by a factor of 2 at each level compared with the reconstructed signals, which have the same number as the original signal.

Wavelet Packet Transform

The WT is actually a subset of a far more versatile transform, the WPT (Cody 1994). Wavelet packets are particular linear combinations of wavelets (Wickerhauser 1994). They form bases that retain many of the orthogonality, smoothness, and localization properties of their parent wavelets.

In DWT, only approximation coefficients are decomposed at every level using high-pass and low-pass filters whereas the detail coefficients, which constitute high-frequency components, are left untouched after every level of decomposition. This may cause problems if any important information is hidden in the higher frequency content of the signal.

The frequency resolution of the decomposition filter may not be fine enough to extract the necessary information from the decomposed component of the signal. On the other hand, the necessary frequency resolution can be achieved by implementing a WPT to decompose a signal further (Goswami and Chan 1999).

In addition, a wavelet packet is represented as $\psi_{j,k}^i(t)$, where i is the modulation parameter, j is the dilation parameter, and k is the translation parameter and is given by

$$\psi_{j,k}^i(t) = 2^{-j/2}\psi^i(2^{-j}t - k) \tag{3.18}$$

here $i = 1, 2..., j^n$ and n is the level of decomposition in the wavelet packet tree. The wavelet $\psi^i(t)$ is obtained by the following recursive relationships:

$$\psi^{2i}(t) = \frac{1}{\sqrt{2}}\sum_{k=-\infty}^{\infty}h(k)\psi^i\left(\frac{t}{2}-k\right) \tag{3.19}$$

$$\psi^{2i+1}(t) = \frac{1}{\sqrt{2}}\sum_{k=-\infty}^{\infty}g(k)\psi^i\left(\frac{t}{2}-k\right) \tag{3.20}$$

where $\psi^i(t)$ is called the mother wavelet and the discrete filters $h(k)$ and $g(k)$ are quadrature mirror filters (QMFs) associated with the scaling function and the mother wavelet (Daubechies 1992).

The wavelet packet coefficients $WPT^i_{j,k}(t)$ corresponding to the signal $f(t)$ can be obtained as

$$WPT^i_{j,k}(t) = \int_{-\infty}^{\infty} f(t)\psi^i_{j,k}(t)dt \tag{3.21}$$

provided the wavelet coefficients satisfy the orthogonality condition.

The wavelet packet component of the original signal (inverse WPT) at level i can be obtained as

$$f^i_{j,k}(t) = \sum_{k=-\infty}^{\infty} WPT^i_{j,k}(t)\psi^i_{j,k}(t)dt \tag{3.22}$$

After performing wavelet packet decomposition up to the jth level, the original signal can be represented as a summation of all wavelet packet components at jth level as

$$f(t) = \sum_{i=1}^{2^j} f^i_j(t) \tag{3.23}$$

Thus, the WPT is similar to the DWT; however, it decomposes the approximate as well as the detail coefficients to achieve an adequate level of frequency resolution, which is not the case with the DWT (Cody 1994).

Yet another offshoot of the WT, namely, the SWT is somewhat similar to the DWT except that the signal under study is never subsampled and instead the filters are upsampled at each level of decomposition. Although the wavelet, in its different forms, are widely being used in numerous applications in science and engineering, data compression can be cited as one of the most significant of all its applications (Provaznik and Kozumplik 1997). Furthermore, the WT are extensively being used in many applications in the field of seismic signal analysis, including noise attenuation studies, stacking of seismic data, spectral decomposition, etc. (Sinha 2002; Victor 2002).

Fast Wavelet Transform

In analogy with FFT algorithm, which computes the discrete Fourier transform (DFT) in a much faster rate, the DWT can also be implemented by a similar algorithm called fast wavelet transform (FWT), which saves considerable computational time in comparison with direct computation of DWT. This is also called the QMF, in which orthogonal decomposition by low-pass and high-pass filters H and G, respectively, takes place at every step in signal processing. In QMF, the decomposition step consists of applying a low (high) pass filter to the signal followed by downsampling and retaining only the even index samples. The reconstruction step consists of upsampling by assuming "0" between adjacent samples followed by filtering and addition to the samples of the next higher level.

If $G = g(n)$ and $H = h(n)$ for a positive integer n, then the reconstruction filters have impulse response $h(n) = h(1 - n)$, and $g(n) = g(1 - n)$. The filters H and G correspond to one step in the wavelet decomposition. Given a discrete signal, $f(k\Delta t)$, with samples of 2^n, at each stage of the WT, the G and H filters are applied to the signal and the filtered output downsampled by two, generating two bands, G and H. The process is then repeated on the H band to generate the next level of decomposition and so on. It is important to note that the wavelet decomposition of a set of discrete samples has exactly the same number of samples as in the original due to the orthogonality of the wavelets. This procedure is referred to as the one-dimensional (1-D) FWT. The inverse FWT can be obtained in a way similar to that of the forward transform by simply reversing the above procedure. However, the order of the impulse responses g and h has to be reversed. The efficiency of the FWT depends on many factors including the type of machine in which it is implemented.

Two-Dimensional Wavelet Transforms

In applications such as image processing and analysis of 2-D geophysical data, we need to have two-dimensional (2-D) WT because the image is mathematically represented by a function of two variables of the form $f(x, y)$. The 2-D WT can be easily generated from its 1-D counterpart. To extend the WT to 2-D, the wavelet function ψ and the scaling function ϕ need to be changed. Although it would be possible to construct wavelets directly in the respective dimension, it is easier to use a separable construction in which the new wavelets are tensor products of 1-D functions (Mallat 1991; Daubechies 1992). If ψ and ϕ form a basis for the 1-D space, then the newly constructed functions will form a basis for the higher dimension as well (Prateepasen and Methong 2003). In the 2-D case, there are three different detail coefficients in each decomposition. A 1-D time–space function $f(t)/f(x)$ results in a 2-D WT, whereas a 2-D function $f(x,y)$ results in a four-dimensional WT (Prateepasen and Methong 2003).

Thus, the 2-D CWT of a real function $f(x,y)$ can be defined as (Zhu et al. 2002)

$$\text{CWT}_f^{\psi}(a,b,s) = \frac{1}{\sqrt{s}} \iint f(x,y)\psi\left(\frac{x-a}{s}, \frac{y-b}{s}\right) dx\, dy \qquad (3.24)$$

and its inverse is given by

$$f(x,y) = \frac{1}{c_\psi} \int\limits_s \int\limits_a \int\limits_b \text{CWT}_f^{\psi}(a,b,s) \frac{1}{s^2} \psi\left(\frac{x-a}{s}, \frac{y-b}{s}\right) da\, db\, ds \qquad (3.25)$$

where a and b represent translation parameters in x and y directions, respectively, and s represents the scaling parameter. For practical applications, the discretized versions of Equations 3.24 and 3.25 need to be used and are called the 2-D DWT, which is presented below.

The 2-D scaling and wavelet functions can be expressed as $\phi(x, y)$ and $\psi(x, y)$, respectively, and are given as

$$\phi(x, y) = \phi(x)\phi(y) \qquad (3.26)$$

$$\psi^h(x, y) = \phi(x)\psi(y) \qquad (3.27)$$

$$\psi^v(x, y) = \psi(x)\phi(y) \qquad (3.28)$$

$$\psi^d(x, y) = \psi(x)\psi(y) \qquad (3.29)$$

where $\psi^h(x, y)$, $\psi^v(x, y)$ and $\psi^d(x, y)$ are wavelets along the directions of horizontal, vertical, and diagonal, respectively. Hence, a 2-D WT can be used to characterize an image by identifying different features of the image at different detail levels (scales) and along different directions such as texture analysis or fingerprint analysis (Kasaei 1998). This is the approach used to extend the Mallat orthogonal wavelet algorithm into two (Mallat 1989) and three dimensions (Meneveau 1991).

The detailed wavelet coefficients along the horizontal, vertical, and diagonal are given by

$$\text{DWT}_f^{\psi^h}(j,k) = \iint f(x,y)\psi_{j,k}^h(x,y)dx\, dy \qquad (3.30)$$

$$\mathrm{DWT}_f^{\psi^v}(j,k) = \iint f(x,y)\psi_{j,k}^v(x,y)\,dx\,dy \qquad (3.31)$$

$$\mathrm{DWT}_f^{\psi^d}(j,k) = \iint f(x,y)\psi_{j,k}^d(x,y)\,dx\,dy \qquad (3.32)$$

where approximation wavelet coefficients are given as

$$\mathrm{DWT}_f^\phi(j,k) = \iint f(x,y)\phi_{j,k}(x,y)\,dx\,dy \qquad (3.33)$$

The detail coefficients given by Equations 3.30, 3.31, and 3.32 as well as the approximation coefficients in Equation 3.33 together constitute the 2-D DWT. Furthermore, from the detail and approximation coefficients, the 2-D inverse DWT can be given as

$$f(x,y) = a_{i_0}(x,y) + \sum_{i=-\infty}^{i_0}\left[d_i^h(x,y) + d_i^v(x,y) + d_i^d(x,y)\right] \qquad (3.34)$$

where the horizontal detail $d_i^h(x,y)$, the vertical detail $d_i^v(x,y)$ and the diagonal detail $d_i^d(x,y)$ of the signal, at level i, are defined as

$$d_i^h(x,y) = \sum_{j=-\infty}^{\infty}\sum_{k=-\infty}^{\infty}\mathrm{DWT}_f^{\psi^h}(i,j,k)\psi_{i,j,k}^h(x,y) \qquad (3.35)$$

$$d_i^v(x,y) = \sum_{j=-\infty}^{\infty}\sum_{k=-\infty}^{\infty}\mathrm{DWT}_f^{\psi^v}(i,j,k)\psi_{i,j,k}^v(x,y) \qquad (3.36)$$

$$d_i^d(x,y) = \sum_{j=-\infty}^{\infty}\sum_{k=-\infty}^{\infty}\mathrm{DWT}_f^{\psi^d}(i,j,k)\psi_{i,j,k}^d(x,y) \qquad (3.37)$$

and the approximation of the signal at level i_0 is

$$a_{i_0} = \sum_{j=-\infty}^{\infty}\sum_{k=-\infty}^{\infty}\mathrm{DWT}_f^\phi(i_0,j,k)\phi_{i_0,j,k}(x,y) \qquad (3.38)$$

The wavelet representation of an image, $f(x,y)$, can be obtained by first applying the 1-D FWT to each row of the image and then to each column of the resultant image, which is known as the pyramid algorithm. That is, the G and H filters are applied to the image in both horizontal and vertical directions. The process is repeated several times as in a 1-D case. This procedure is referred to as the 2-D FWT, which can easily be extended to multidimensions. However, it has two drawbacks; the DWT lacks shift invariance, which means that small shifts in input signal can cause big changes in the energy distribution of the wavelet coefficients. Secondly, the DWT has poor directional selectivity for diagonal features. Despite the inherent weakness of 2-D WT, its applications are varied and wide-ranging in different fields, particularly in image processing, resulting in an understanding of fluid velocity patterns from images (Anderson and Hongyan 1995). In geophysics, the power of wavelets has been used for the analysis of nonstationary processes that contain multiscale features, the detection of singularities, the analysis of transient phenomena, fractal, and multifractal processes, and for signal analysis and compression, among others (Praveen 1997).

Second-Generation Wavelets

Traditionally, wavelet functions $\psi_{\tau,s}(t)$ are defined as translates and dilates of one particular function, the mother wavelet $\psi(t)$. We refer to these as first-generation wavelets (FGW). Because translation and dilation in the frequency domain are linear algebraic operations, these wavelets remain invariant over the entire signal or image to be analyzed. Furthermore, many signals and images do not possess the fractal self-similarity property, and multiresolution analysis of these signals using linear wavelet dilates of FGW, therefore, are not suitable. Generalizations of this type, in which wavelets need not and in fact cannot be translated and dilated, of one or a few templates are called second-generation wavelets (SGW; Sweldens 1997). Before the generalization of the SGW, the following properties of FGW should be preserved.

P1: The wavelets and their duals are local in space and frequency. Some wavelets are even compactly supported.

P2: The frequency localization follows from the smoothness of the wavelets (decay toward high frequencies) and the fact that they have vanishing polynomial moments (decay toward low frequencies).

P3: Wavelets fit into the framework of multiresolution analysis. This leads to the FWT, which allows us to pass between the function f and its wavelet coefficients in linear time.

The applications that illustrate the need for generalizations in FGWs may be noted here.

G1: Although FGWs provided bases for functions defined on R^n applications such as data segmentation and the solution of partial differential and integral equations, in general, domains require wavelets that are defined on arbitrary, possibly nonsmooth domains of R^n.

G2: Diagonalization of differential forms, analysis on curves and surfaces, and weighted approximation require a basis adapted to weighted measures; however, FGWs typically provide bases only for spaces with translation-invariant measures.

G3: Many real-life problems require algorithms adapted to irregularly sampled data, whereas FGWs imply a regular sampling of the data.

There are various techniques to construct SGWs. A generalization of FGWs to the settings G1 through G3, while preserving the properties P1 through P4, is needed. The main limitation is that the FGW works well for infinite or periodic signals but it should be modified for use in a bounded domain because, in many applications, the domain of interest is not infinite and signals are not periodic. Furthermore, even 1-D signals are often not sampled regularly. However, this can also be done in the spatial domain by using a new approach called *the lifting scheme*. The basic idea is to first split a signal into its even and odd samples. That is, split an original signal to even indexed samples x_e and odd indexed samples x_o. Recall that wavelets start with the key observation that digital signals are highly correlated, and the correlation structure is local. So, the idea here is to predict the odd signal from the even part. Typically, these two sets are closely correlated. Thus, it is only natural that given one set, for example, x_o, one can build a good predictor P for the other set, for example, x_e. However, the predictor need not be exact, so we need to record the difference or detail d as (Cohen et al. 1992)

$$d = x_o - P(x_e) \qquad (3.39)$$

If P is an accurate predictor, then d will be a very sparse set. The lifting scheme replaces x_0 by d, thus achieving data reduction. The detail signal d is a WT coefficient that is a measure of how much the local portion of the original signal fails to be linear. It's the output of the high-pass filter. Let us consider a simple linear predictor for an odd indexed sample by averaging two adjacent even indexed neighbors. The detail d can be recorded as simply the average of its two even neighbors; the detail coefficient then is

$$d_k = x_{2k+1} - (x_{2k+2} + x_{2k})/2 \qquad (3.40)$$

where k is a positive integer

From this we see that if the original signal is locally linear, the detail coefficient is zero. The operation of computing a prediction and recording the detail is called the lifting step. The subsampling action resulting from the splitting step usually cannot guarantee adequate spectral separation and may produce aliasing in the two polyphase sets x_e and x_0. The final step in the lifting scheme is to update x_e by replacing it with an aliasing-free smoother set s, ready for the next lower resolution lifting stage. In the original lifting scheme, the update operator U is a linear combination of the detail signal d:

$$s = x_e + U(d) \qquad (3.41)$$

Again, this step is trivially invertible: given (s, d) we can recover x_e as

$$x_e = s - U(d) \qquad (3.42)$$

and then x_0 can be recovered as explained earlier. This illustrates one of the built-in features of lifting: no matter how P and U are chosen, the scheme is always invertible and thus leads to critically sampled perfect reconstruction filter banks.

An update operator that restores the correct running average, and therefore reduces aliasing, is given by

$$s_k + x_{2k} + (d_{k-1} + d_k)/4 \qquad (3.43)$$

where k is a positive integer.

Note that the idea of using spatial wavelet constructions for building SGW has been proposed by several researchers: the lifting scheme is inspired by the work of Donoho (1992) and Lounsbery et al. (1997). Donoho (1992) shows how to build wavelets from interpolating scaling functions, whereas Lounsbery et al. (1997) built a multiresolution analysis of surfaces using a technique that is algebraically the same as lifting. Dahmen and collaborators worked on stable completions of multiscale transforms, a setting similar to SGWs independent of lifting (Carnicer et al. 1996; Dahmen et al. 1994). Again, independently, the development of a general multiresolution approximation framework based on spatial prediction was in vogue (Harten 1996). Dahmen and Micchelli (1993) proposed the construction of compactly supported wavelets that generate complementary spaces in a multiresolution analysis of univariate, irregularly knotted splines.

The SGWT has a number of advantages over the classic WT in that it is quicker to compute (by a factor of 2) and it can be used to generate a multiresolution analysis that does not fit a uniform grid. Using a priori information, the grid can be designed to allow the best analysis of the signal to be made. The transform can be modified locally while preserving invertibility; it can even adapt, to some extent, to the transformed signal.

Applications of WT to Potential Fields

Geophysical signal processing techniques attempt to extract hidden information from a data set measured on the surface of the earth. In the case of potential field (PF) measurements, the goal is to learn about the distribution of the material properties in the subsurface aside from exactly locating the subsurface resources. Also, the interpretation of PF data covers the lateral and vertical extensions of sources including their depth, width, dip, and estimating their tonnage, etc. Processing and interpretation of various geophysical data can be realized by the applications of integral transforms. Many of these techniques include the Fourier and Hilbert transforms (Sundararajan et al. 1998), and their offshoots. In this direction, the application of WT in the processing and interpretation of geophysical data has been gaining importance recently (Lyrio et al. 2004; Cooper 2006; Xu et al. 2009; Sailhac et al. 2009).

Because the space frequency localization characteristics of signals can be better understood in WT than in FT, the use of WT in PF studies is justified. The main physical principle of wavelets is to observe that the PF caused by a homogeneous source such as a monopole or dipole, is homogeneous too. The idea behind the use of homogenous source is that an extended geological body causing a PF anomaly may be replaced by a small number of equivalent point sources. The equivalent sources associated with an extended body will depend on the scale (i.e., the dilation "a"). At small scales, the equivalent sources will be localized on the edges of the geological source, whereas at large scales, the equivalent sources will be more global.

Another important aspect of assuming homogenous causative sources is that when upward continued at a level $z_0 + z_0'$, the potential is dilated and with a rescaled version of PF measured at the reference surface z_0 above the source. It may be derived from this that the altitude offset of the upward continuation (z_0') is equivalent to the dilation a when the analyzing wavelet belongs to the Poisson family. This property ($z_0' = a$) is greatly useful in the estimation of source depth when the PF data is subjected to CWT.

In most of the WT applications in science and engineering, we use the traditional wavelets like Haar, Marlet, Mexican hat, Daubechies, etc. Among the hundreds of available mother wavelets, the question of which mother wavelet to use is indeed a big question to begin with, but with experience, one can overcome this. For many applications in different fields, *db5* is preferred. These wavelets do find applications in geophysics except perhaps in the interpretation of PF data and hence the necessity for the Poisson wavelet family.

The Green function of the Poisson equation and its spatial derivatives lead to a family of wavelets specifically tailored to PF. The Poisson family of wavelets can be generated by applying homogenous Fourier multipliers of arbitrary degree $\gamma \geq 1$ to the Poisson semigroup kernel which, in 1-D, can be defined as (Hassina and Dominique 2006)

$$P(x) = C/(1 + |x|^2) \tag{3.44}$$

where C is a normalizing constant. The Fourier multiplier from the point of view of PF theory is equivalent to computing vertical and horizontal derivatives.

The Poisson kernel family has only limited, but definite, use in geosciences for PF applications. It is only Poisson kernel wavelets that enhanced our understanding of the sources responsible for PF signals. The Poisson kernel family wavelets enable the depth calculation of the source of the measured PF signal. The general equation of the horizontal derivative of order n of the Poisson kernel family, $H_n(u)$(Moreau et al. 1997; Saracco et al. 2004), is given as

$$H_n(u) = (2\pi u)^n \exp(-2\pi|u|) \tag{3.45}$$

where u is the wave number of the spatial variable x and n being the order of the derivative.

Similarly, through the Hilbert transform of the horizontal derivative, the general equation of the vertical derivative of order n of the Poisson family, $V_n(u)$ is given as (Mauri et al. 2011)

$$V_n(u) = -2\pi|u|(2\pi iu)^{n-1}\exp(-2\pi|u|) \tag{3.46}$$

PF signals analyzed by these wavelets $V_n(u)$ and $H_n(u)$ allow for the estimation of source depth. The wavelets $[V_n(u)$ and $H_n(u)$; $n = 1$–5] are very similar in shape, but each has its own center frequency and hence will react slightly differently to the same signal and noise. Although the law of derivative order can determine the most appropriate choice of the wavelet, there is no objective way to decide which wavelet will best characterize the given signal. Interpreters of PF data can create many such wavelets to the Poisson kernel on a trial and error basis. Wavelets derived from the Poisson kernel are based on a gravity anomaly from a point source. This makes it possible to achieve the scaling of the wavelet in CWT with an upward continuation operation.

The 2-D wavelet belonging to the Poisson semigroup class is defined as (Sailhac et al. 2009)

$$\psi \frac{1}{x}(b) = -\frac{3x}{2\pi(x^2 + y^2 + 1)^{5/2}} \tag{3.47}$$

It is mostly used in the source depth estimation from PF anomalies.

Denoising seeks to eliminate or minimize noise and preserve signals. In this direction, wavelet-based denoising is superior to traditional Fourier domain filters in many applications of PF data. Frequency-based filters have the adverse effect of smoothing out high-frequency signals. At times, denoising and smoothing are commonly considered as synonymous; however, they differ in many applications including PF data. Denoising aims to remove/minimize noise while preserving the signal. On the other hand, smoothing removes/minimizes all high-frequency content in the data irrespective of whether it belongs to signal or noise. Wavelet filtering overcomes this effect and therefore ideal for denoising. Although the choice of mother wavelet is a bit problematic in applications, by and large, Daubechies wavelets of different order are reported to be fairly sound in denoising of PF field data. An efficient automatic denoising of gravity gradiometry data was illustrated by Lyrio et al. (2004), using the wavelet filtering technique.

Wavelet denoising is a three-step process wherein we apply WT W to the noisy data g as (Lyrio et al. 2004)

$$w = Wg \qquad (3.48)$$

where w is the wavelet coefficient that is thresholded as

$$v = \delta_\lambda^H(w) \qquad (3.49)$$

where $\delta_\lambda^H(w)$ is a hard threshold function with threshold λ defined as $\delta_\lambda^H(w) = 0$ if $|w| \leq \lambda$ and $\delta_\lambda^H(w) = w$ if $|w| \geq \lambda$.

Finally, the estimated signal g_e is obtained as the inverse WT, that is, W^{-1} to the modified set of coefficients v as

$$g_e = w^{-1}v \qquad (3.50)$$

Denoising by thresholding requires the selection of thresholds and of the initial scales where thresholding begins. Soft threshold introduces more bias; however, it is efficient in interactive denoising (Lyrio et al. 2004).

Generally, the measured PF geophysical data (anomalies) after all corrections represent the combined response of both large deep-seated structures and small shallow-depth sources known as Bouger gravity anomaly. Before interpretation, it is necessary to separate the corresponding regional (low frequency—long WL due to basins/geosyncline) and residual (high frequency—short WL due to salt dome/anticline) components for an accurate determination of depth and other parameters of the source. There are several methods that are in vogue; wavelet-based methods, although exciting, are yet to be practiced. In the recent past, the WT has been widely used for processing and interpretation of gravity data due to the compact property of multiscale analysis and has turned out to be an important and reliable tool for anomaly separation (Xu et al. 2009).

Bouger gravity anomaly = regional anomaly + residual anomaly

Alternatively, the applications of WT to PF data interpretation uses wavelets derived from the Poisson kernel, which is based on the gravity anomaly from a point source. This makes it possible to achieve the scaling of the wavelet in CWT with an upward continuation operation. For example, the gravity response due to a buried 2-D cylinder and its first and second horizontal derivatives are given as (Cooper 2006)

$$g(x) = \frac{Gmz}{(x^2 + z^2)} \tag{3.51}$$

$$\frac{dg(x)}{dx} = \frac{-2Gmzx}{(x^2 + z^2)^2} \tag{3.52}$$

$$\frac{d^2 g(x)}{dx^3} = \frac{2Gmz(3x^2 - z^2)}{(x^3 + z^3)^3} \tag{3.53}$$

Here, the derivatives (Equations 3.52 and 3.53) are suitable as mother wavelets whereas the anomaly itself is not because Equation 3.44 does not have a zero mean, which is a prerequisite for any function to be a wavelet. In all these equations, G and m may be ignored, and z can be set to one for simplicity. It is important to note that the data to which wavelet analysis is to be applied needs to be differentiated to the same degree as the mother wavelet.

For CWT analysis of the vertical derivative of the gravity anomaly, the wavelet used must be horizontal derivatives of the vertical derivative of the gravity field given in Equation 3.51.

$$\frac{d^2 g(x)}{dzdx} = \frac{-2Gmx(x^2 - 3z^2)}{(x^3 + z^3)^3} \tag{3.54}$$

$$\frac{d^2 g(x)}{dzdx^2} = \frac{6Gm(x^4 + z^4 - 6x^2 z^2)}{(x^2 + z^2)^4} \tag{3.55}$$

Equations 3.54 and 3.55 are the first and second horizontal derivatives of the vertical derivative of the gravity effect of the cylinder (Equation 3.51). Despite this rigorous mathematical exercise, the WTs are not amenable for the extraction of source parameters quantitatively, but may produce reliable results qualitatively.

Therefore, wavelet analysis may have practical applications, particularly in PF (gravity and magnetic) data in the following areas:

(a) Denoising

(b) Regional residual separation

(c) Source depth estimation

But in the case of estimation of depth to the source, it is not comparable with the traditional Fourier/Hartley spectral analysis or analytic signal approach or even any other integral transform methods for the reasons that the WT-based method is neither straightforward nor does it ensure simple solutions. Generally, it is qualitative or semiquantitative with added complexities. In certain cases, it is not even an independent tool for the interpretation of geophysical data and needs added information from other tools like Radon, FTs, etc. However, in the case of denoising and regional residual separation of PF data are found to be effective and perhaps it still needs to be simplified.

Overall, the processing and interpretation of PF anomalies is not unified in wavelet domain and it is case-to-case based and therefore it needs to be done a lot more for practical applications specifically in source depth estimation.

References

Abramovich, F., Bailey, T.C., and Sapatinas, T. 2000. Wavelet analysis and its statistical applications. *The Statistician Journal of the Royal Statistical Society, Series D* 49, 1–29.

Allen, J., and Rabiner, L. 1977. A unified approach to short-time Fourier analysis and synthesis. *Proceedings of the IEEE* 65, 1558–1564.

Anderson, W.L., and Hongyan, D. 1995. Two dimensional wavelet transform and application to holographic particle velocimetry. *Applied Optics* 34, 244–259.

Bracewell, R.N. 1983. The discrete Hartley transform. *Journal of the Optical Society of America* 73, 1832–1835.

Burrus, C.S., Gopinath, R.A., and Guo, H. 1998. *Introduction to Wavelets and the Wavelet Transforms. A Primer*. Prentice Hall, Upper Saddle River, NJ.

Carnicer, J.M., Dahmen, W., and Pena, J.M. 1996. Local decompositions of refinable spaces. *Applied and Computational Harmonic Analysis* 3, 127–153.

Cody, M.A. 1994. The wavelet packet transform. *Dr. Dobb's Journal* 19, 44–46, 50–54.

Cohen, A., Daubechies, I., and Feauveau, J. 1992. Biorthogonal bases of compactly supported wavelets. *Communications on Pure and Applied Mathematics* 45, 485–560.

Cooley, J.W., and Tukey, J.W. 1965. An algorithm for the machine calculation of complex Fourier series. *Mathematics of Computation* 19, 297–301.

Cooper, G.R.J. 2006. Interpreting potential fields data using continuous wavelet transforms and their horizontal derivatives. *Computers and Geosciences* 32, 984–992.

Dahmen, W., and Micchelli, C.A. 1993. Banded matrices with banded inverses. II: Locally finite decompositions of spline spaces. *Constructive Approximation* 9, 263–281.

Dahmen, W., Prossdorf, S., and Schneider, R. 1994. Multiscale methods for pseudo-differential equations on smooth manifolds. In *Proceedings of the International Conference on Wavelets: Theory, Algorithms, and Applications*. 385–424. Academic Press.

Daubechies, I. 1992. *Ten Lectures on Wavelets*. Society for Industrial and Applied Mathematics. Philadelphia, PA.

Donoho, D.L. 1992. *Interpolating Wavelet Transforms*. Preprint, Department of Statistics, Stanford University, San Francisco, CA.

Foufola-Georgoiu, E., and Kumar, P. 1994. Wavelets in geophysics. In *Wavelet Analysis and its Applications*, 4. Academic Press, Los Angeles, CA.

Gabor, D. 1946. Theory of communication. *Journal of the Institution of Electrical Engineers*, 93, 429–441.

Goupillaud, P., Grossmann, A., and Morlet, J. 1984. Cycle-octave and related transforms in seismic signal analysis. *Geoexploration* 23, 85–105.

Goswami, J., and Chan, A. 1999. *Fundamentals of Wavelets: Theory, Algorithms and Applications*. Wiley-Interscience, Hoboken, NJ.

Grossmann, A., and Morlet, J. 1984. Decomposition of Hardy functions into square integrable wavelets of constant shape. *SIAM Journal on Mathematical Analysis* 15, 723–736.

Harten, A. 1996. Multiresolution representation of data: A general framework. *SIAM Journal on Numerical Analysis* 33, 1205–1256.

Hassina, D., and Dominique, G. 2006. Identification of sources of potential fields with the continuous wavelet transform: Two dimensional ridgelet analysis. *Journal of Geophysical Research* 111, B07104, 11.

Kasaei, S. 1998. Fingerprint Analysis Using Wavelet Transform with Application to Compression and Feature Extraction. PhD Thesis, Queensland University of Technology.

Lounsbery, M., DeRose, T.D., and Warren, J. 1997. Multiresolution surfaces of arbitrary topological type. *ACM Transactions on Graphics* 16, 34–73.

Lyrio, J.C.S.O., Luis, T., and Li, Y. 2004. Efficient automatic denoising of gravity gradiometry data. *Geophysics* 68, 772–782.

Mallat, S.G. 1989. A theory for multiresolution signal decomposition: The wavelet representation. *IEEE Transactions on Pattern Analysis and Machine Intelligence* 11, 674–693.

Mallat, S.G. 1991. Multiresolution approximations and wavelets orthonormal bases. *Transactions of the American Mathematical Society* 315, 334–351.

Mauri, G., Jones, G.W., and Saracco, G. 2011. MWTmat-application of multiscale wavelet tomography on potential fields. *Computers and Geosciences* 37, 1825–1835.

Meneveau, C. 1991. Analysis of turbulence in the orthonormal wavelet representation. *Journal of Fluid Mechanics* 232, 469–520.

Moreau, F., Gibert, D., Holschneider, M., and Saracco, G. 1997. Wavelet analysis of potential fields. *Inverse Problem* 13, 165–178.

Morlet, J. 1981. Sampling theory and wave propagation. In *Proceedings, 51st Annual Meeting, Society of Exploration Geophysicists, Los Angeles*.

Morlet, J. 1983. *Sampling Theory and Wave Propagation*. 1, 233–261. Springer, New York.

Polikar, R., 1996. *The Wavelet Tutorial Part III*. Rowan Univ. College of Engineering, Web Submission.

Prateepasen, A., and Methong, W. 2003. Characterization of wire rope defects from magnetic flux leakage signals. *Thammasat International Journal of Science and Technology*, 8, 54–63.

Praven, K. 1997. Wavelet analysis for geophysical applications. *Reviews of Geophysics* 35, 385–412.

Provaznik, I., and Kozumplik, J. 1997. Wavelet transform in electrocardiography-data compression. *International Journal of Medical Informatics* 45, 111–128.

Qian, S., and Chen, D. 1996. *Joint Time-Frequency Analysis*. Prentice Hall, NJ.

Sailhac, P., Gibert, D., and Bookerbout, H. 2009. The theory of continuous wavelet transform in the interpretation of potential fields—a review. *Geophysical Prospecting* 57, 517–525.

Saracco, G., Labazuy, P., and Moreau, F. 2004. Localization of self-potential sources in volcano-electric effect with complex continuous wavelet transform and electrical tomography methods for an active volcano. *Geophysical Research Letters* 31(12), L12610. doi:10.1029/2004GL019554.

Sinha, S. 2002. Time-Frequency Localization with Wavelet Transform and Its Application in Seismic Data Analysis. Master's Thesis, University of Oklahoma.

Strang, G., and Nguyen, T. 1996. *Wavelets and Filter Banks*. Wellesley-Cambridge Press, Boston.

Sundararajan, N. 1995. 2-D Hartley transforms. *Geophysics* 60, 262–267.

Sundararajan, N. 1997. Fourier and Hartley transforms—a mathematical twin. *Indian Journal of Pure and Applied Mathematics* 28, 1361–1365.

Sundararajan, N., Srinivasa Rao, P., and Sunitha, V. 1998. An analytical method to interpret SP anomalies due to 2-D inclined sheets. *Geophysics* 63, 1551–1555.

Sweldens, W. 1997. The lifting scheme: A construction of second generation wavelets. *SIAM Journal on Mathematical Analysis* 29, 511–546.

Torrence, C., and Compo, G. 1998. A practical guide to wavelet analysis. *Bulletin of the American Meteorological Society*, 79, 61–78.

Victor, I. 2002. Seismic Signal Enhancement Using Time-Frequency Transforms. Master's Thesis, University of Calgary.

Wickerhauser, V. 1994. *Adapted Wavelet Analysis from Theory to Software*. 213–214, 237, 273–274, 387. AK Peters, Boston.

Xu, Y., Hao, T., Li, Z., Duan, Q., and Zhang, L. 2009. Residual gravity anomaly separation using wavelet transform and spectrum analysis. *Journal of Geophysics and Engineering* 6, 279–287.

Zhu, S., Li, T., and Ogihara, M. 2002. An algorithm for non-distance based clustering in high dimensional spaces. In *Proceedings of 4th International Conference, number 2454 in LNCS, Aix-en-Provence, France*, 52–62. Springer, New York.

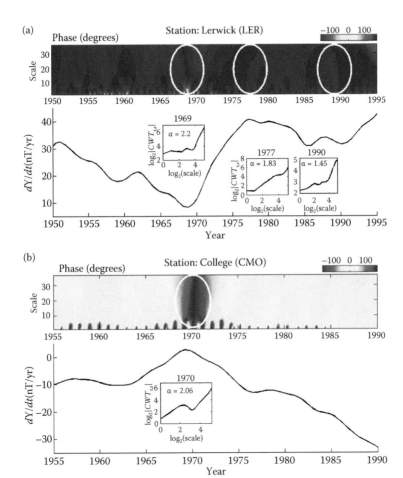

FIGURE 7.7
Phase scalogram plots together with secular variation plots and the ridge function plots related to the 1969–1970 event corresponding to (a) LER and (b) CMO observatories of polar region.

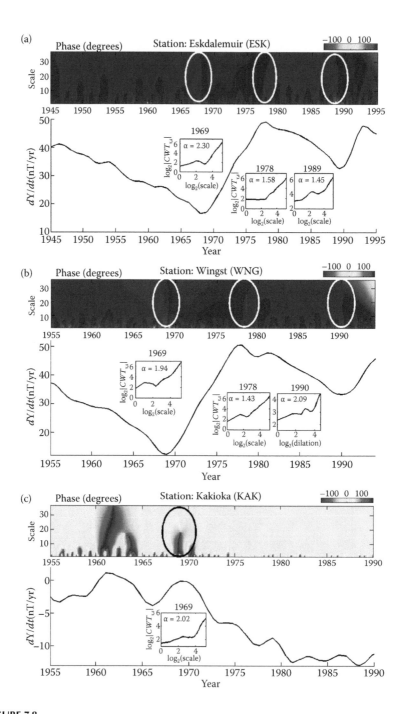

FIGURE 7.8
Phase scalogram plots together with secular variation plots and the ridge function plots related to the global jerk events corresponding to (a) ESK, (b) WNG, and (c) KAK observatories of the high-latitude to mid-latitude regions.

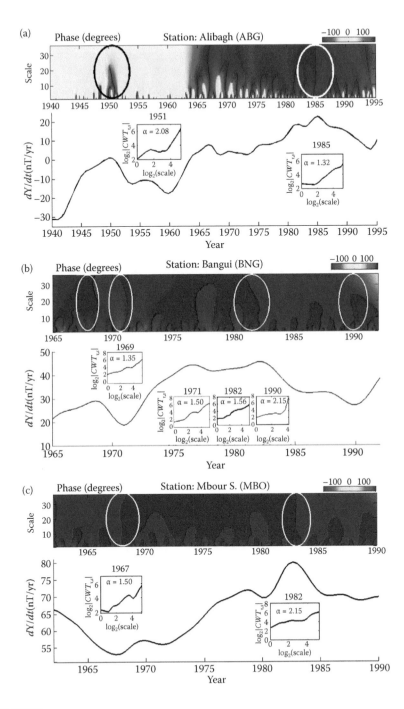

FIGURE 7.9
Phase scalogram plots together with secular variation plots and the ridge function plots related to the global and local jerk events corresponding to (a) ABG, (b) BNG, and (c) MBO observatories of the mid-latitude to equatorial regions.

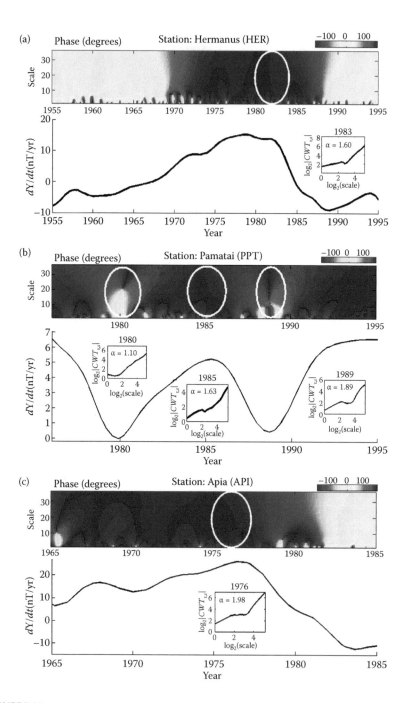

FIGURE 7.10
Phase scalogram plots together with secular variation plots and the ridge function plots related to the local jerk events corresponding to (a) HER, (b) PPT, and (c) API observatories of the southern hemisphere region.

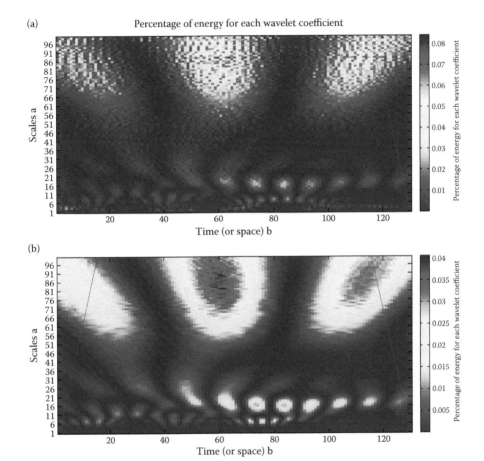

FIGURE 9.3
(a) Wavelet spectrum of rainfall data without smoothening; (b) wavelet spectrum of rainfall data after smoothening.

4

Multiscale Processing: A Boon for Self-Similar Data, Data Compression, Singularities, and Noise Removal

Ratnesh S. Sengar, Venkateswararao Cherukuri, Arpit Agarwal, and Vikram M. Gadre

CONTENTS

Introduction 118
Dyadic Multiresolution Analysis.. 119
Axioms of Dyadic MRA 119
    Ladder Axiom 119
    Axiom of Perfect Reconstruction 121
    The Intersection of All the Subspaces is the Trivial Subspace {0}........... 121
    If a Function $x(t) \in V_0$ then, $x(2^i t) \in V_i$, $i \in \mathbb{Z}$ 121
    Similarly, if $x(t) \in V_0$ then $x(t - n) \in V_0$, $n \in \mathbb{Z}$ 122
    Axiom of Orthogonal Basis... 122
Theorem of MRA.. 122
Wavelet-Based Data Compression... 124
Fractals and Self-Similar Functions 131
    Fractals 131
Self-Similar Functions... 133
Wavelet Transform of Self-Similar Functions... 134
Application of Self-Similar Functions 135
Singularities and Noise Removal.. 138
    Singularities... 138
Wavelets in Singularity Detection.. 139
    Singularities in a Signal without Noise.. 139
    Singularities in Noisy Signals.. 140
    Wavelets in Denoising 141
    Thresholding Techniques for Signal Denoising 141
Two Case Studies of Singularity Detection 143
    One-Dimensional Singularity in ECG Signals 143
    Two-Dimensional Singularities in Proteomic Images.............................. 148
Acknowledgments... 153
References.. 153

Introduction

As we have seen in previous chapters, the wavelet transform has allowed scientists and engineers to analyze the time-varying and transient phenomena of a signal. The continuous wavelet transform (CWT) is used to measure the similarity between a signal and the analyzing function wavelet at various scales and locations. The CWT compares the signal to the shifted and compressed or stretched versions of the wavelet, also called translations and dilations of the wavelet, respectively. The discrete version of CWT is used to actually calculate the wavelet transform. This discrete version is called the discrete wavelet transform (DWT). Dyadic sampling of the time–frequency plane results in a very efficient algorithm for calculating the DWT. Dyadic multiresolution analysis (MRA) is one of the techniques that help us study how to analyze functions in space $\mathcal{L}^2(\mathbb{R})$ at different scales. The different scales at which the functions are analyzed are related by powers of two.

In this chapter, we are going to see how and why wavelets are used in processing data across scales. The axioms of dyadic MRA are briefly discussed in this chapter with a different perspective, bringing out the inherent self-similar properties in the wavelet functions, which are efficiently used for multiscale processing. Before looking at the different perspectives of these axioms, an overview is given about the different contexts in which wavelets are efficiently used in processing data across scales exploiting the interscale relationships. Three such contexts are discussed in this chapter. A small overview is discussed below and a detailed explanation follows in the subsequent sections.

One of the contexts in which we use the multiscale character of wavelets is for data compression. In this situation, we demonstrate how wavelets exploit the data across different scales, or subbands, to form tree type data structures in the wavelet domain. These tree type data structures, which are formed by exploiting the interscale relations between data, are efficiently processed by some algorithms such as set partitioning in hierarchical trees (SPIHT) [1] and embedded zero wavelet (EZW) [2] to achieve good compression. In this context, the SPIHT algorithm is explained using an example to demonstrate how wavelets exploit interscale relationships in given data.

The other context in which we use wavelets for processing data across scales is for processing self-similar functions, often called *fractals*. Self-similar functions or fractals are a class of functions in which the data across different scales of the function look similar. These functions will be explained mathematically in later sections. Because the data across the scales for these functions are similar and because wavelets are an effective tool for processing data across scales, we study how the wavelet transform of these functions look like. Then, an example is used to demonstrate how wavelets are efficiently used in studying a special class of self-similar functions known as fractional Brownian motions. We see how a fractional Brownian motion is characterized by exploiting a persistent scale-to-scale relationship in the wavelet data.

In the third part of the chapter, we will study singularities and noise removal. Noise removal is an important preprocessing step for analyzing any real signal or detecting singularities. First, we present a brief overview of the different types of noise present in signals. Then, we will study the role of wavelets in denoising and different denoising techniques involving the wavelet transform. Following this, we will look into the importance of wavelets for singularity detection. Singularity detection for one-dimensional (1-D) and two-dimensional (2-D) signals will be explained using the examples of electrocardiogram (ECG) signals and proteomic images.

For all the contexts discussed above, the wavelet transform emerges as a powerful tool for identifying and exploiting interscale relationships. This is because these situations reveal critical information as we go across scales. Tools other than wavelets or MRA are not as effective in studying that critical information. Therefore, this chapter aims to reveal different methods using wavelets in the contexts discussed above. Before that, we see why the axioms, from which dyadic wavelets emerge, justify the use of dyadic wavelets as an efficient tool for processing data across scales. For this purpose, a small introduction to the axioms of dyadic MRA is given.

Dyadic Multiresolution Analysis

In dyadic MRA, we analyze functions in the space of square integrable functions, $\mathcal{L}^2(\mathbb{R})$, at different scales. In dyadic MRA, the scale is discretized by powers of two and the time axis is covered by discrete steps, corresponding to the given scale. The basic theory of dyadic MRA helps in understanding why wavelets are actually used for processing data across scales. Therefore, a different perspective of the axioms of dyadic MRA and theorem of MRA is presented below. To make it easier to appreciate this, we use the Haar MRA to illustrate the nuances of the discussion.

Axioms of Dyadic MRA

Ladder Axiom

The name "ladder axiom" emerges from the way subspaces are organized. Let $V_0, V_1, \ldots\ldots, V_n$ be a ladder of subspaces that belong to $\mathcal{L}^2(\mathbb{R})$ in which each subspace V_i, $i \in \mathbb{Z}$ contains functions that are piecewise constant in the intervals $[2^{-i}l, 2^{-i}(l + 1)]$, $l \in \mathbb{Z}$. Then,

$$\ldots \subset V_{-2} \subset V_{-1} \subset V_0 \subset V_1 \subset V_2 \ldots$$

Intuitively, a function which is piecewise constant in the intervals $[2^{-i}l, 2^{-i}(l + 1)]$, $l \in \mathbb{Z}$ is also piecewise constant in the intervals $[2^{-(i+1)}l, 2^{-(i+1)}(l + 1)]$, $l \in \mathbb{Z}$. To span this entire subspace V_i, a set of linearly independent functions that belong to V_i are used. These are called the basis functions for V_i. From the above axiom, we can say that the basis functions of V_{i+1} can be used to represent the functions in V_i.

As the above axiom is interpreted, it is observed that as a function in a lower subspace is contracted, it becomes a part of a higher subspace and vice versa, as illustrated in the next subsection. It implies that the above subspaces are self-similar in nature. Hence, the bases of the above subspaces are also self-similar. As we can see, each subspace contains the information of a signal corresponding to that "scale." Therefore, the property of a subspace embedded in another subspace is exclusively used for analyzing singularities and self-similar functions. A singularity is a point in a function in which it blows up or degenerates in some particular way, such as lack of differentiability or discontinuity. This property of singularities propagates through scales. This is because the information obtained at one scale is embedded in another scale. So, a trend is observed at singular points across each scale. By observing this trend, we can detect singularities in the given data. We will discuss singularities in more detail in later sections. The ladder axiom also brings out the fact that bases of MRA are self-similar in nature. This, in turn, helps in analyzing the self-similar functions effectively.

Figure 4.1 shows functions F_0, F_1, F_2, which belong to subspaces V_0, V_1, and V_2, respectively. Rectangles in the figure show the maximum resolution possible in each subspace (1 in V_0, 1/2 in V_1, and 1/4 in V_2), that is, the scale of the wavelet function used in them. The values a, b, c, d, e, f, and g depend on the signal amplitude. For the functions F_0 and F_1, if we put the condition

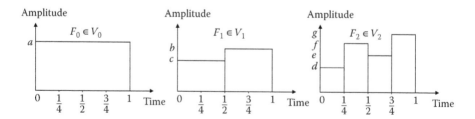

FIGURE 4.1
Illustration of a ladder axiom.

$b = c$, then $F_0 \equiv F_1$. This means F_0 is a special case of an element in subspace V_1. Similarly, if the condition is $d = f$ and $e = g$, then $F_1 \equiv F_2$, which means F_1 is a special case of an element in subspace V_2. Also, if $d = e = f = g$, then F_0 is a special case of an element in subspace V_2. From the above inferences, we can conclude that $V_0 \subset V_1 \subset V_2$, and this can be extended to the ladder axiom, which encompasses all the subspaces V_k.

This axiom, combined with other axioms, lead us to the theorem of MRA. The result of the theorem of MRA and the above axiom is exploited in many applications in one way or another.

Axiom of Perfect Reconstruction

When a closed union of all the V_i s is taken, we reach the $\mathcal{L}^2(\mathbb{R})$ space.

$$\overline{\bigcup_{i \in \mathbb{Z}} V_i} = \mathcal{L}^2(\mathbb{R})$$

Intuitively we can say that if we take union of all the functions which are piecewise constant at different scales of powers of 2, we can go arbitrarily close to a function in $\mathcal{L}^2(\mathbb{R})$. This axiom reveals that a function in $\mathcal{L}^2(\mathbb{R})$ can be reconstructed almost perfectly if the information across each scale or in each subspace is preserved.

The Intersection of All the Subspaces is the Trivial Subspace {0}

This is because a function cannot have finite energy and be piecewise constant with an infinite support. This axiom also reveals that a function in $\mathcal{L}^2(\mathbb{R})$ has significantly less information at scales that have large support, and as the length of the timescale reaches infinity, the content of the signal approaches zero. The mathematical equation is shown below.

$$\bigcap_{i \in \mathbb{Z}} V_i = 0$$

If a Function $x(t) \in V_0$ then, $x(2^i t) \in V_i$, $i \in \mathbb{Z}$

Intuitively, we can say that if a function belongs to V_0, then the timescaled version of the same function belongs to its respective V_i. This axiom is very useful in analyzing self-similar functions, as self-similar functions also exhibit a similar type of property, which will be discussed in later sections. This axiom implies that when a function is contracted or expanded by a factor of 2, the function moves into different subspaces in the same ladder.

Similarly, if $x(t) \in V_0$ then $x(t - n) \in V_0$, $n \in \mathbb{Z}$

It can be generalized as, if $x(t) \in V_i$, then $x(t - n2^i) \in V_i$, $n \in \mathbb{Z}$, that is, if a function belongs to V_i, then the same function translated by a discrete step also belongs to V_i. The discrete step permissible depends on the scale at which we work.

Axiom of Orthogonal Basis

There exists a function $\phi(t)$ and its integer translates, which form an orthogonal basis for V_0. Then, by using Axioms 4 and 5, we can establish the orthogonal basis for the rest of the V_is, that is, $\phi(2^i t)$ and its appropriate discrete translates form the basis for V_i. This axiom lays down the basic foundation in decomposing the signal into different scales.

These subspaces can also be thought of as those that give some approximate information of a function at each timescale. Therefore, while we move from one subspace V_i to the next subspace V_{i+1}, we require some incremental information about the function. The theorem of MRA helps us establish results about this incremental information. From the above results, we can actually demonstrate a relation between the ladder of subspaces. From Axiom 1 and Axiom 6, we can establish the following result:

$$\phi(t) = \sum_{n \in \mathbb{Z}} h(n)\phi(2t - n), \ h(n) \text{ is a discrete sequence.}$$

This relation is valid because any function in V_0 can be represented with the basis of V_1. It can be seen that $\phi(t)$ can be expressed as a linear combination of its contracted and translated versions $\phi(2t - n)$, $n \in \mathbb{Z}$. This property is exhibited by self-similar functions. Hence, the bases of these V_i subspaces are self-similar. These bases, which are themselves self-similar, are very useful in representing self-similar data.

Theorem of MRA

Given the above axioms of a dyadic MRA, there exists a function $\psi(t) \in \mathcal{L}^2(\mathbb{R})$ and $\psi(t) \in V_1$ such that $\{\psi(2^m t - n)\}_{m \in \mathbb{Z}, n \in \mathbb{Z}}$ forms an orthogonal basis for $\mathcal{L}^2(\mathbb{R})$.

The proof of this theorem has been explained in a few video lectures [3]. The space spanned by $\{\psi(2^m t - n)\}_{m \in \mathbb{Z}, n \in \mathbb{Z}}$ is denoted by W_m. These subspaces can be thought as incremental subspaces, which provide the incremental information of a function while moving from one scale to another.

While establishing the proof of this theorem, we can arrive at several auxiliary results. We know that,

$$\bigcup_{i \in \mathbb{Z}} V_i = \mathcal{L}^2(\mathbb{R}),$$ (4.1)

Also,

$$V_{m+1} = V_m \oplus W_m$$ (4.2)

Using Equations 4.1 and 4.2, we can form a chain equation resulting in the following equation:

$$\mathcal{L}^2(\mathbb{R}) = V_0 \oplus \left(\oplus_{n=0}^{n=\infty} W_n \right)$$ (4.3)

Using the above results, any function $f(t)$ in $\mathcal{L}^2(\mathbb{R})$ can be represented as follows:

$$f(t) = \sum_{n=-\infty}^{n=\infty} a_n \phi(t-n) + \sum_{n=-\infty}^{n=\infty} \sum_{m=0}^{\infty} b_{m,n} \psi(2^m t - n),$$ (4.4)

where $a_n = \int_{<t>} f(t) \phi(t-n) dt$ and $b_{m,n} = 2^{m/2} \int_{<t>} f(t) \psi(2^m t - n) dt$

The representation of the signal as shown in Equation 4.4 is called the dyadic wavelet transform. The coefficients a_n and $b_{m,n}$ provide a lot of information of a signal at a particular scale and in a specific region of time, in the frequency band corresponding to that scale. The spectral content of $\phi(t)$ is concentrated around the zero frequency, whereas that of $\psi(t)$ is concentrated around a nonzero frequency. Therefore, the functions $\phi(t)$ and $\psi(t)$ are approximately low-pass and band-pass functions. Particularly, $\psi(t)$ at different scales acts as the impulse response of the band-pass filter with different pass band frequencies. So, if we consider a particular $b_{m,n}$ it can loosely be said to give information about the signal in the interval $[n2^{-m}, (n + 1)2^{-m}]$ and information about spectral content in the corresponding pass band frequencies of $\psi(2^m t)$. When a function is decomposed in this manner, information about that function in a specific region of time and in a particular frequency band corresponding to the given scale is obtained. This is the reason why wavelets have become an efficient tool for processing data across scales. This way of representing data helps in many applications of signal processing.

Three such contexts have already been briefly mentioned. One of those applications is data compression. In data compression, we need to represent the signal by encoding with fewer bits than that required to represent the data in the original signal. When this encoded bit stream is decoded, we should be able to reconstruct the signal to a good degree of accuracy, with a good compression ratio. In the next section, we show how the representation of a signal (as in Equation 4.4) helps in efficient encoding of the data, by exploiting the interscale relationship.

Wavelet-Based Data Compression

In general, the data in signals and images is highly correlated. In other words, there is a high amount of redundancy in images and signals. So, our primary goal in image or signal compression is to decorrelate data. This can be done by representing the data in terms of a different basis, such that the data in this new representation is uncorrelated. There are many bases that represent the signal in such a way that the data becomes uncorrelated. This way of representing a signal in terms of different bases can be done by applying a suitable transform on the signal. For example, applying the Fourier transform on the signal transforms the signal in terms of the Fourier basis, that is, representing in terms of complex phasors. The discrete cosine transform (DCT) represents a discrete signal in terms of discrete cosine basis elements. In both of the above transforms, the basis elements are orthonormal. Also, in both domains, we analyze the amplitude or energy of the signal at different frequencies. However, we fail to analyze the properties of the signal simultaneously in both the frequency and time domains. Although global information can be obtained from the transformed data, localized information cannot be obtained using these two transforms. Localized information or information at a particular scale and around a region of time can be obtained from the data by decomposing the signal into the wavelet basis as shown in Equation 4.4.

The representation of the signal as shown in Equation 4.4 is called the wavelet transform of the signal. In this domain, we will be able to analyze the signal at different frequencies around a region of time. In general, we deal with discrete signals. Without loss of generality, we can assume that the discrete signal belongs to the space V_i. Then, we can decompose the signal into subsequent lower subspaces, that is, to $V_0, W_0, W_1, ..., W_{i-1}$ by using the DWT. Once the wavelet transform is completed, we use some efficient encoders for this data. The encoders that are used will efficiently exploit the properties of wavelets. When a given data is decomposed into wavelet coefficients, the wavelet coefficients at similar spatial locations in different subbands can be grouped together to form tree type data structures across the entire data, known as hierarchical trees. The encoders which are going to be discussed exploit this hierarchical relation between different subbands of the data obtained after the DWT. Two of these encoders are SPIHT [1] and the

EZW [2]. In the following section, an example of image compression by using DWT along with a SPIHT encoder is shown. Also, we show how the SPIHT algorithm uses the relation between different subbands of the data efficiently.

Example:

Consider an image of resolution $n \times n$ (where n is a power of 2). Because the image is a discrete signal, we can, without loss of generality, consider it as a piecewise constant function in the intervals of $[n2^{-2}, (n + 1)2^{-2}]$, that is, the image belongs to V_2. Let the image be represented by a function $F(x, y)$. First, we need to decompose this image into different subbands. Because an image is composed of 2-D data, the decomposition has to be done in both horizontal and vertical directions. The way the image is decomposed is shown in Figure 4.2, and that same image, after wavelet transformation, is shown in Figure 4.3. The image is decomposed according to the following equation:

$$V_2 = V_0 \oplus W_0 \oplus W_1 \tag{4.5}$$

Once the image is decomposed into different subbands, we need to efficiently encode this data. In general, the significant pixels after the wavelet transform are sent to the encoder. We need to identify the pixels that are significant. In a bit level encoder, we test the significance of each pixel for all the bit planes. For example, if an image is represented with 16 bits per pixel, the 16th bit is used for representing the sign of the pixel and the remaining 15 bits represent the magnitude of the pixel.

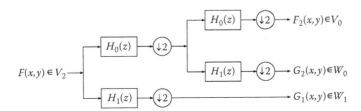

FIGURE 4.2
Image is decomposed into different subbands of frequencies using a two-band filter bank.

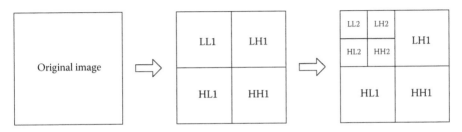

FIGURE 4.3
Decomposition of image into different subbands.

Therefore, while encoding the first bit plane, check whether the 15th bit of a pixel is significant or not, that is, if the 15th bit of pixel is 1, we say the pixel is significant for that given bit plane, else it is insignificant.

This significance test is required for every pixel and for each bit plane. If the image is large, it takes lot of time to check the significance of each pixel of the image for each bit plane. Instead of checking for each pixel, we need to form a set of data so that the significance test can be carried on this set. Let us denote this set of pixels by Γ. It is important that the group of pixels that are in a set should share some common properties among themselves. The SPIHT algorithm groups the data from different subbands into a set. In the 1-D case, when a signal is decomposed into V_1 and W_1 from V_2, the size of the signal in V_1 and W_1 subspaces is half that of the original signal in V_2. Intuitively, two consecutive points of the signal in subspace V_2 share some properties with a single coefficient at a similar spatial location in subspaces V_1 and W_1. Because the image is essentially 2-D data, one pixel of $F_2(x, y)$ (i.e., image in V_1) shares properties with four pixels of the original image [i.e., $F(x, y)$] at a similar spatial location. The main question is, which group of pixels forms such a set. For an image, once we apply the wavelet transform, we get one approximate subspace and three incremental subspaces (i.e., one V_i and three W_is). For example, in Figure 4.3, the LL1 subband corresponds to the approximate subspace and the remaining three subbands correspond to incremental subspaces. The LL1 subband is further subdivided into one approximate subspace and three incremental subspaces. Figure 4.4 shows a two-level wavelet transform for the benchmark "Lena" image. We can observe that a similar type of pattern is repeated across different scales as shown in Figure 4.4. This is because the process of the wavelet-based decomposition is self-similar. The SPIHT encoder exploits this self-similar process of decomposing the signal. In a later section, we see how wavelets are used in detecting singularities. There, when the wavelet transform is observed at different scales, the points where singularities are present are local maxima at all the scales, and the wavelet transform at the remaining points gets suppressed in all the scales. This property is exploited by the SPIHT algorithm. Thus, the pixels located at similar

(a)   (b)   (c)

FIGURE 4.4
DWT of Lena image: (a) original image, (b) first-level decomposition, and (c) second-level decomposition.

spatial locations at different subbands are grouped together to form a tree type data structure, in which the pixels located at coarser scales are called parents and the pixels located at finer scales are called children.

For example, the LL2 subband is called the coarser scale and the remaining subbands are called finer scales. The SPIHT algorithm defines a hierarchical structure in which a coefficient at the coarser scale is called a parent, and the coefficients located at same spatial locations in the next finer scales of similar orientation are called children. By doing so, if a singularity is present in the coarser scale, then this singularity is propagated to all scales at a similar spatial location. Because all these pixels are grouped into a family, the whole set of pixels in that family are at a local maxima at their respective scales. With a high probability, these pixels become significant. Suppose the pixel that we consider at the coarser scale is not a singularity. Then the magnitude of the wavelet transform at a similar spatial location in different subbands is low. So, with high probability, all these pixels are insignificant. The two cases discussed above reveal that structural significance or insignificance is propagated across scales. Therefore, by observing the parents at coarser scales, we can draw certain conclusions about the entire "family" of that pixel. For a given parent, there exist four children in the next finer scale. For each coefficient in the LL2 subband, barring a few coefficients, there exist children and grandchildren in subsequent finer scales.

This hierarchical representation of the parent–child relation across different scales is called a spatial-orientation tree. Figure 4.5 shows how

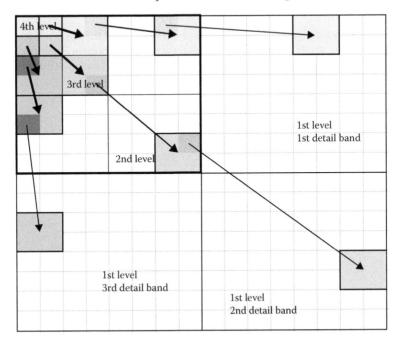

FIGURE 4.5
A hierarchical structure of coefficients in spatial orientation tree.

a typical spatial-orientation tree is formed across different scales. Once a tree is formed for a pixel in the LL2 subband, the following sets are defined based on the properties we discussed above.

Let the coordinate of a pixel be (i, j). We shall use the term *node* for a pixel in the sequel. Then, $O(i, j)$: the set of four "children" of node (i, j), also called offspring for node (i, j). The coordinates of the four children of the pixel (i, j) are

$[(2i, 2j), (2i + 1, 2j), (2i, 2j + 1),(2i + 1, 2j + 1)]$
$D(i, j)$: all the "descendants" of the node (i, j) including the children
$L(i, j) = D(i, j) - O(i, j)$, all the descendants of the node (i, j) other
　　than the children.
H: All pixels in the LL2 subband.

Once these three sets are defined, we define three lists as follows:

LIP: list of insignificant pixels. All the pixels that are insignificant
　　in a given sorting pass are moved to LIP. It is initialized to H.
LSP: list of significant pixels. All the pixels that are significant in
　　a given sorting pass are moved to LSP. It is initialized to the
　　null set.
LIS: list of insignificant sets. LIS is divided into type A and type B,
　　in which type A contains all the insignificant $D(i, j)$s and type B
　　contains all the insignificant $L(i, j)$s. LIS is initialized to all $D(i, j)$s,
　　that is, initialize all the (i, j)s $\in H$ as type A entry.

A set is said to be insignificant if the maximum value in the set is insignificant. So, an operator $S_n(i, j)$ is defined such that if a pixel is significant, $S_n(i, j) = 1$, else $S_n(i, j) = 0$. Before starting the algorithm, initialize $n = \log_2(\max_{i,j}(|C_{i,j}|))$, where $C_{i,j}$ is the value of pixel (i, j).

Algorithm:

1. Sorting pass
　1.1 For each entry in LIP output $S_n(i, j)$. If $S_n(i, j) = 1$, then move the
　　　pixel (i, j) to LSP and output the sign of the pixel (i, j)
　1.2 For each entry in LIS do:
　　　1.2.1 If the entry is of type A, then output $S_n(D(i, j))$
　　　1.2.2 If $S_n(D(i, j)) = 1$, then for each $(k, l) \in O(i, j)$
　　　　　– Output $S_n(k, l)$. If $S_n(k, l) = 1$, then add (k, l) to the end of
　　　　　　LSP and output the sign of the pixel (k, l), else add (k, l) to
　　　　　　the end of LIP
　　　　　– If $L(i, j) = \emptyset$, then move (i, j) to the end of LIS, as an entry of
　　　　　　type B, and go to 1.2.3, else remove (i, j) from LIS
　　　1.2.3 If the entry is of type B, then output $S_n(L(i, j))$
　　　　　– If $S_n(L(i, j)) = 1$, then add each $(k, l) \in O(i, j)$ to the end of the
　　　　　　LIS as an entry of type A and remove (i, j) from LIS

2. Refinement pass: for each entry (i, j) included in LSP, except those added in the current sorting pass, output the nth most significant bit of $|C_{i,j}|$

3. Decrement n by 1 and go to step 1

Said and Pearlman have pioneered the algorithm and one can refer to their seminal article for details [1]. In this way, the image is encoded and sent to the decoder. The decoder is symmetric to the encoder. It can be understood by simply replacing *output* in the algorithm with *input*. Once the information is decoded, the inverse wavelet transform is computed and the original image is reconstructed. The beauty of the SPIHT algorithm is that we can achieve both lossless and lossy compression by truncating step 3 in accordance with user convenience. Because the image is reconstructed at the receiver, if the user finds the quality of the image obtained to be acceptable, the algorithm can be stopped at step 3 for a desired value of n. If the user feels lossless compression is needed, then the algorithm continues until the value of n reaches 0.

In this section, we have shown the advantage of representing an image using wavelet basis. This advantage is obtained as the process of wavelet-based decomposition of a signal is self-similar in nature. For demonstration, a two-level wavelet decomposition of an 8 × 8 image is taken and shown in Figure 4.6, in which we illustrate how the hierarchical trees are formed. The first corner box represents the LL2 subband and the remaining subbands can be understood from the previous discussion. It can be

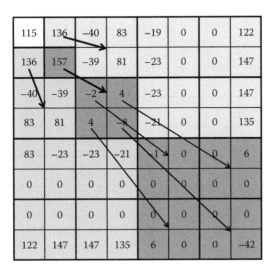

FIGURE 4.6
Formation of hierarchical trees for an 8 × 8 image.

seen that the pixels at similar locations at different subbands are grouped together. Once these trees are formed, the next step is the first sorting pass. Because the image taken is represented using 8 bits per pixel, in the first sorting pass, all the pixels that have their most significant bit as 1 are sent. The data after the first sorting pass is shown in Figure 4.7. It can be observed that the box containing the pixels {−2, 4, 4, −8} are insignificant and the pixels that belong to the family of these pixels are also insignificant in the next finer scale. Structural insignificance is observed in this case. As we can see, the data that passes through the first sorting is not entirely reconstructed. This data is refined in the second sorting pass, as discussed in the algorithm. The data after the third sorting pass (and from the second refinement pass) is shown in Figure 4.8. This procedure can be continued until eight passes, or we can truncate in between, when the quality of the image obtained is acceptable.

From these figures, it can be observed how the formation of tree-type data structures is efficiently used in data compression. In this section, we have seen how structural insignificance or significance across the scales is exploited and efficiently used for data compression. In this application, we have observed how the data across scales shares common properties when the wavelet transform is applied on a signal. In the next section, we shall study the relation between data across scales when the wavelet transform is applied on a special class of functions called fractals or self-similar functions and investigate how wavelets help in characterizing these signals.

0	128	0	0	0	0	0	0
128	128	0	0	0	0	0	128
0	0	0	0	0	0	0	128
0	0	0	0	0	0	0	128
0	0	0	0	0	0	0	0
0	0	0	0	0	0	0	0
0	0	0	0	0	0	0	0
0	128	128	128	0	0	0	0

FIGURE 4.7
Data after first sorting pass.

96	128	−32	64	0	0	0	96
128	128	−32	64	0	0	0	128
−32	−32	0	0	0	0	0	128
64	64	0	0	0	0	0	128
0	0	0	0	0	0	0	0
0	0	0	0	0	0	0	0
0	0	0	0	0	0	0	0
96	128	128	128	0	0	0	−32

FIGURE 4.8
Data after third sorting pass and second refinement pass.

Fractals and Self-Similar Functions

Fractals

From our basic understanding of geometry, we say that a line has one dimension, a square has two dimensions, a cube has three dimensions, and so on. In all these cases, we observe that the dimensions of the objects described are whole numbers. The objects we just mentioned are regular objects. There is one school of thought which says that most of the objects that are present in nature are not regular. Therefore, we need a method for finding the dimension for these irregular objects. Hausdorff defined a formula for finding the dimension of any object, whether it is regular or irregular as [4]

$$\lim_{\varepsilon \to 0} \frac{(\ln N(\varepsilon))}{\ln\left(\dfrac{1}{\varepsilon}\right)} \tag{4.6}$$

where $N(\varepsilon)$ is the number of square boxes of length ε needed to enclose the object. The dimension described above can also be called the box-counting dimension. By observing the above formula, we can say that the dimension of an object can be a fraction. The geometrical objects that have fractional dimension are called as "fractals." For regular objects such as lines, squares, cubes, and so on, the above dimension turns out to be the dimension of space in which the object is embedded. For a fractal, the dimension lies between $n - 1$ and n, where n is the dimension of the space in which the fractal object is embedded. An example of a fractal object is the Cantor set [4]. As shown in Figure 4.9, the middle 1/3rd part of each line segment is removed and a union of the remaining line segments is taken. This set can be obtained by applying the two functions $f_1(x) = \dfrac{x}{3}$ and $f_2(x) = \dfrac{x}{3} + \dfrac{2}{3}$ on line segment [0, 1] and taking their union in every iteration. Because this object is embedded in a 1-D space, we expect the dimension of the object to lie between 0 and 1. Taking $\varepsilon = \dfrac{1}{3}$ and substituting in Equation 4.6, we get the dimension of the Cantor set as ln2/ln3, which lies between 0 and 1.

Similarly, we take another example of a fractal, which is embedded in 2-D space, known as the Koch curve. As shown in Figure 4.10, initially an equilateral triangle is taken and each line segment of the triangle is divided into three equal parts. Finally, an equilateral triangle is formed with the middle part of the line segment as one of the sides. Then, the middle part of the line segment is removed. This process is done iteratively, and finally, the Koch curve is obtained in the "limit." In Figure 4.10, the portion that is circled in

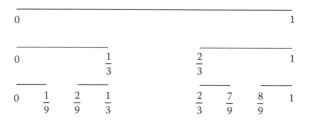

FIGURE 4.9
Cantor set.

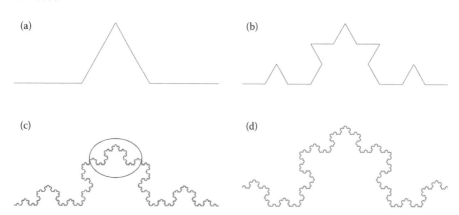

FIGURE 4.10
Evolution of Koch curve: (a) initiator, (b) curve after first iteration, (c) final Koch curve after multiple iterations, and (d) zoomed portion circled in (c).

part c is shown in part d. It can be observed that the entire image in part c looks similar to the image shown in a zoomed version in part d. The dimension of the Koch curve turns out to be $\dfrac{\ln 4}{\ln 3}$, which lies between 1 and 2, illustrating this idea of dimension again. Similarly, there exist many other fractals in nature. As we observe in the above two cases, an initiator is taken and a set of functions are applied on it. Their union is taken iteratively to form a fractal set or fractal curve. This entire procedure is called an iterative function scheme (IFS). More information about fractals can be obtained from the book by Stephane Mallat [5].

Self-similarity is a typical property of fractals. In the next section, we will further discuss self-similar functions.

Self-Similar Functions

A continuous function f with compact support S is said to be self-similar if there exist disjoint sets $S_1, \ldots, S_k$ such that the graph of f restricted to each S_i is an affine transformation of f. This can be mathematically written as follows [5]:

$$\forall t \in S_i, f(t) = c + w_i f[s_i(t - l_i)] \tag{4.7}$$

Because fractals are obtained from applying an affine transformation on a set recursively, they exhibit self-similarity. The best examples of self-similar functions are the scaling functions of any MRA. Figure 4.11 demonstrates the self-similar property for the scaling function of the Daubechies wavelet emerging from filter length 4 (DB2). It can be observed that the scaling

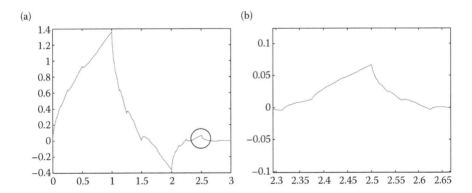

FIGURE 4.11
Illustration of self-similar nature of the DB2 wavelet scaling function: (a) scaling function for the DB2 wavelet, and (b) the zoomed portion of the scaling function for the DB2 wavelet.

function of DB2 shown in Figure 4.11a, and the circled portion of the same scaling function, shown zoomed in Figure 4.11b, look similar. This property of self-similarity is the basis used for MRA that helps us in dealing with self-similar functions effectively.

Wavelet Transform of Self-Similar Functions

As a consequence of this self-similarity in basis elements, it is interesting to know how the wavelet transform of self-similar functions looks. The wavelet transform of a self-similar function computed at different scales is also self-similar. This can be shown by taking a special class of self-similar functions that obey the relation shown in Equation 4.8 below

$$x(st) = s^H x(t) \tag{4.8}$$

These functions are also called fractional Brownian motions. Any function can be expressed in terms of the orthonormal wavelet $\psi(t)$ as follows:

$$x(t) = \sum_{m \in \mathbb{Z}} \sum_{n \in \mathbb{Z}} x_{m,n} 2^{m/2} \psi_{m,n}(t), \text{ where } x_{m,n} = \int_{<t>} 2^{m/2} x(t)\psi(2^m t - n)\,dt.$$

By changing the scales of integration $2^m t = \alpha$ and using the relation in Equation 4.8, we obtain $x_{m,n} = 2^{-(H+1/2)m} x_{0,n}$.

This expression proves that the wavelet transform of a self-similar function is also self-similar. This result can also be proved, in general, for a self-similar function [6]. For example, consider the self-similar function shown in Figure 4.12. The first-level wavelet transform of the function is shown in

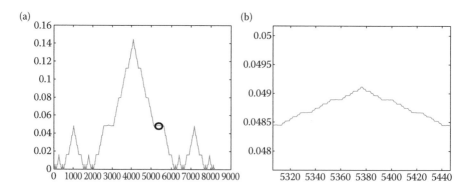

FIGURE 4.12
An example of a self-similar function (a) Von Koch curve, (b) zoomed portion circled in (a).

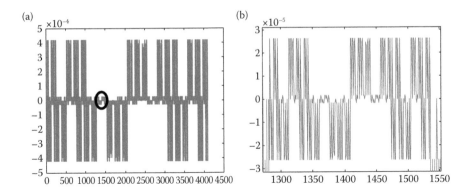

FIGURE 4.13
Wavelet transform of the signal shown in Figure 4.12a and corresponding self-similarity.

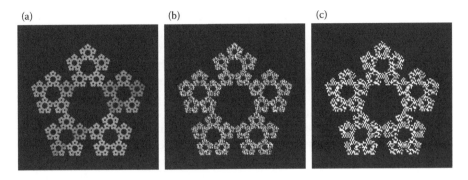

FIGURE 4.14
Illustration of wavelet transform for a fractal image: (a) fractal image, (b) wavelet transform of the image at scale 1, and (c) wavelet transform for the image at scale 2.

Figure 4.13. In part a, a certain portion of the image is circled and shown in part b. It can be observed that the shape of the wavelet transform of the circled portion is similar to that of the entire image shown in part a. The same holds for fractal images. Figure 4.14 demonstrates the wavelet transform of a fractal image. It can be observed that the wavelet transform of the image at different scales is also self-similar. In the next section, we are going to study fractional Brownian motions and apply the property that we just described.

Application of Self-Similar Functions

Self-similarity has been exploited in many areas such as modeling traffic in broadband networks, fluctuations in the stock market, natural images, and so on. All the applications described above exhibit a special property. Their autocorrelation function remains scale invariant. These processes are also called $1/f$ processes, meaning that their power spectral density function is inversely proportional to f^γ, where γ is a parameter that determines how fast the autocovariance of the function decays. These processes are also called fractional Brownian motions. If $B_H(t)$ is any $1/f$ process or a fractional Brownian motion, then the following relation holds [7,8]

$$B_H(st) \equiv s^H B_H(t) \tag{4.9}$$

where $2H + 1 = \gamma$ and $0 < H < 1$. This parameter H is called the Hurst parameter. The "equality" or equivalence in the above equation is in a statistical sense, that is, their probability density functions are equal but the actual values at a given time may not be equal. The Hurst parameter is also referred to, as the "index of long-range dependence." If $0.5 < H < 1$, then the data has a long-term positive correlation, meaning that there is a high chance that a higher value is followed by another higher value, and that this trend follows for a long range of values. On the other hand, for $0 < H < 0.5$, the data has a long-term switching tendency between adjacent pairs, that is, a higher value is highly likely to be followed by a smaller value and vice versa. So, the Hurst parameter plays an important role in characterizing the given fractional Brownian motion. The estimation of the Hurst parameter becomes critical in analyzing these processes. Because these processes are statistically self-similar and as wavelets are effective in studying a function at different scales, wavelets also become an effective tool for estimating the Hurst parameter. In this section, we briefly describe a method proposed by Miduri, Wirnell, and other researchers [7,8], using wavelets for estimating the Hurst parameter by relating data at different scales. Any function $x(t)$ can be represented as

$$x(t) = \sum_{m \in Z} \sum_{n \in Z} x_{m,n} 2^{m/2} \psi_{m,n}(t) \tag{4.10}$$

where $\psi(t)$ is the wavelet basis for a given MRA. Now, for a given scale m, calculate the energy of coefficients at that scale as

$$\Gamma_m = \frac{1}{n_m} \sum_{n \in Z} |x_{m,n}|^2 \tag{4.11}$$

Γ_m is now the energy of the wavelet coefficients at scale m, n_m is the number of coefficients at scale m. If $x(t)$ is a fractional Brownian motion, then the following relation between different scales of wavelet data holds true [7,8]:

$$\Gamma_m = 2^{m(2H+1)}\Gamma_0 \tag{4.12}$$

Using this relation, if we plot a graph of log Γ_m versus m and then find the slope of the best-fitting straight line in the sense of minimum mean squared error (MMSE), we get γ, from which we can find H. In this way, we can analyze fractional Brownian motion using wavelets. We demonstrate with an example below.

Example:

In this section, we provide an example to illustrate an estimation of the Hurst parameter from a fractional Brownian motion. Computer-generated fractional Brownian motion with a Hurst parameter of 0.7 is shown in Figure 4.15a. As discussed above, the wavelet coefficients are calculated for scales of 1 to 25. Then according to Equation 4.11, the energy of the wavelet coefficients at each scale is obtained. The logarithm with base 2 for energy of the wavelet coefficients at each scale is shown in Figure 4.15b. The dotted line in the graph shown in Figure 4.15b represents the best linear fit for the original graph. The slope of the linear curve according to Equation 4.12 is $2H + 1$.

From Figure 4.15b, the slope of the best linear fit that minimizes the least squares error is 2.3 in which the estimated H is approximately 0.65, which is close to 0.7. This is how the Hurst parameter for a given fractional Brownian motion signal is estimated. In the two contexts discussed until now, we have seen how data across scales share common properties and how the

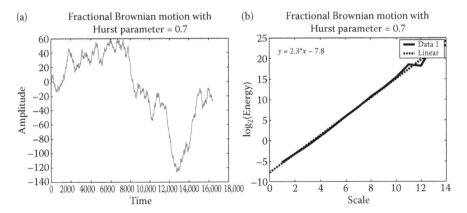

FIGURE 4.15
Scale versus energy of wavelet coefficients: (a) the signal taken for analysis, (b) scale versus $\log_2$ (energy).

data across scales are related. In the next section, we will see another application of wavelets in detecting singularities and in noise removal. In these applications, we use some mathematical tools on the data obtained, after the wavelet transform is taken. Also, we see how singularities can be detected by multiplying the adjacent bands of wavelet transform.

Singularities and Noise Removal

Singularities

A singularity is, in general, a point at which the signal blows up or becomes degenerate in some particular way, such as a loss of differentiability. Singularities and irregular structures generally carry most important information in any signal. In 2-D signals, normally termed images, irregular structures such as sharp changes in intensity provide the contour location and help in recognition. For other signals such as ECGs and sound pressure waves, the interesting information lies in the transients like peaks and troughs, all of which constitute singularities. Also, it is equally important to study the irregularities in any signal, to study its deviation from ideality to detect several abnormalities.

Different types of singularities are shown in Figure 4.16.

Before wavelets became an often used signal processing tool, the Fourier transform was the sole mathematical tool available to analyze singularities. The Fourier transform indicates the global overall regularity of a signal but fails to provide the spatial location of q singularity. Due to this constraint, the short-time Fourier transform (STFT), which could provide the spatial location of singularities, was introduced. However, STFT could not provide good resolution in both time and frequency simultaneously. There had to be a trade-off between time resolution and frequency resolution. The wavelet transform was a carefully meditated approach to deal with these diverse issues [9]. The wavelet transform breaks the signal into several building blocks, which are well localized in both space and frequency simultaneously.

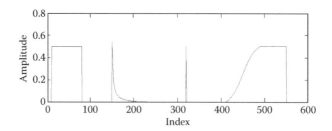

FIGURE 4.16
Types of singularities.

Wavelets in Singularity Detection

Singularities in a Signal without Noise

Stephane Mallat and Wen Liang Hwang explained the notion of singularity detection in a signal by decomposition over a wavelet function, which is the derivative of a zero phase, low-pass smoothing function [10]. They further found that the information in a signal is mainly represented by the wavelet transform modulus maxima (WTMM) and can be used to locate the singularities in any signal, whether at 1-D or 2-D. One of the main features of the wavelet transform is that it retains and, in fact, emphasizes the points of singularity of the original signal. To locate these singularities, an appropriate threshold can be applied on the subbands. We can get an idea of how the WTMM method works by using the simple example in Figure 4.17.

In the above example, singularities are located at (x = 30, 60, 90) in the original signal (without noise). It is visible from the figure that the locations of these singularities are retained across the scales. If we take the modulus maxima of the wavelet transform at multiple scales, the exact locations of singularities can be obtained. Now, we will see what happens in signals contaminated by noise.

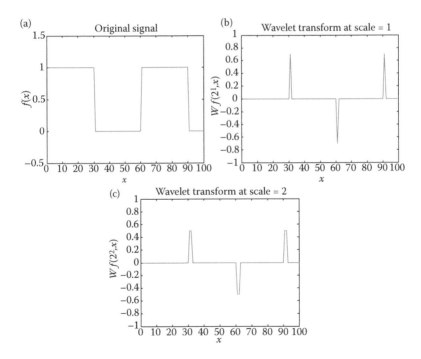

FIGURE 4.17
Retention of singularities across the scales.

Singularities in Noisy Signals

The main theme of this chapter is to exploit the correlation of wavelet information across scales. This correlation is very important in terms of removing the noise from a signal. In the wavelet transform of a noisy signal, the singularities of the original signal and some part of the noise are retained at each scale. A. Pizurica and her colleagues exploited this interscale information successfully to classify the wavelet coefficients [11]. If a coefficient has a smaller magnitude at a coarser scale, then its descendants at finer scales are likely to be small. Conversely, if coefficients corresponding to features are of a larger magnitude at finer scale, its parents at a coarser scale are likely to be large. However, for coefficients caused by noise, the magnitudes will always decay along the scales as shown in Figure 4.18. As one can see, the amplitude

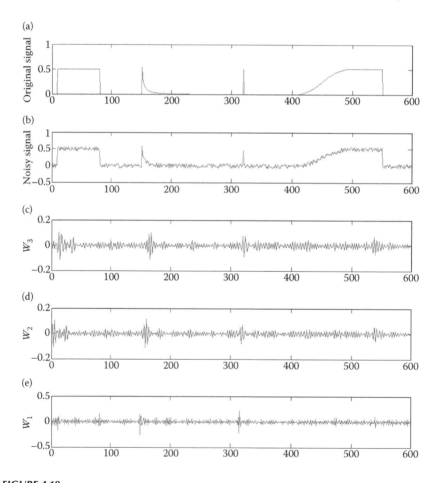

FIGURE 4.18
Illustration of noise removal: (a) original signal, (b) signal with additive random noise, (c) detail coefficients at scale 1, (d) detail coefficients at scale 2, and (e) detail coefficients at scale 3.

of the coefficients corresponding to singularities keeps increasing from W1 to W3, and those corresponding to noise keep decreasing, giving only the signal singularities at scale 3.

We now explore the use of wavelets in denoising.

Wavelets in Denoising

In the nineteenth century, Fourier laid the foundations of frequency analysis of signals. It proved highly significant, but with time, the attention of researchers moved from frequency analysis to scale analysis, when it became clear that only frequency analysis is less effective in dealing with noise. This was because the Fourier transforms only gave the global information of any signal without any localization. The "smooth" and "rough" parts of a signal occur in different "time zones." We need to identify these different zones to denoise effectively. Noise generally constitutes the rough part of the signal or often, more technically, the high-frequency part in any signal. Removing noise does not necessarily mean just smoothing the signal. Denoising removes the unnecessary high-frequency part of the signal while retaining the complete spectrum of the original signal. The Fourier transform helps us find the high-frequency content in the signal but fails to provide the "position" of the same. The Fourier transform can be used to denoise signals in which there is a minimum overlap between the signal and the noise in the frequency domain. However, this approach cannot separate the noise where noise overlaps the Fourier spectra of the signal. Therefore, to denoise effectively, we need to look at both domains to "separate" signal and noise. Also, we need to operate between coefficients across scales to strengthen the signal and weaken noise.

Because the wavelet transform helps us extract the information in the time as well as in the frequency domains simultaneously, it is suitable for dealing with nonstationary signals. Stationary noise can be removed using more conventional techniques, but it is difficult to remove nonstationary noise, such as that which occurs in proteomic images using those techniques. The wavelet transform fragments the spectrum by peeling off a band of frequency at each level while maintaining spatial localization. This idea is clearly shown in Figure 4.19. Figure 4.19a shows how the frequency bands are peeled off at every level of decomposition. To illustrate this, we applied the wavelet transform on the famous Lena image at up to 12 levels. Figure 4.19d shows the diagonal details at levels 0, 4, 8, and 12. It can be seen that as the scale increases, the center frequency associated with the content decreases. Scale 0 shows sharp boundaries in the images whereas scales 8 and 12 have lower frequencies and show smoother boundaries as compared with scale 0 and 4.

Thresholding Techniques for Signal Denoising

Nonlinear thresholding is one of the most simple and yet effective methods to remove noise from the signal. Donoho [12] showed that the universal

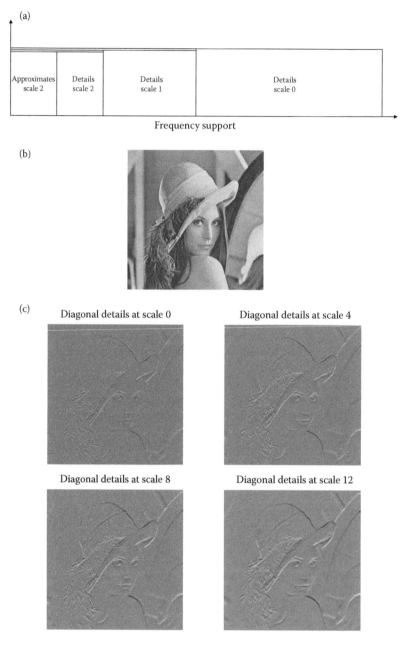

FIGURE 4.19

Frequency contents at different levels shows (a) frequency bands that are peeled off at every level of decomposition, (b) original image with size 512 × 512 pixels, and (c) diagonal details at different scales.

threshold represented by $t = \sigma\sqrt{2\log N}$ (σ is the standard deviation of the white noise and N the sample length) is asymptotically optimal in the minimax sense. Donoho [12] improved his work by introducing the subband adaptive SURE threshold. Then, Chang et al. [13] introduced Bayes shrink, another subband-dependent threshold, the results of which are better than the SURE threshold. However, these are all soft thresholds, which means that the input is shrunk by an amount t. These methods of noise removal are explained in more detail in the relevant research articles [12–14].

After observing the wavelet transform of noisy signals at different scales, Xu et al. [15] and Sadler et al. [16] worked on sharpening the important structures (while weakening noise) by multiplying the adjacent wavelet scales. Both Xu and Sadler used the dyadic wavelet constructed by Mallat and Zhong [17], which is the compactly supported quadratic spline function that approximates the first derivative of the Gaussian.

The thresholding approach does not exploit the inter scale dependencies. However, the approaches described in the previous paragraphs were also that used by of Bao and Zhang [18], who multiplied the adjacent bands of the wavelet transform to make use of the interscale dependencies, as shown in Figure 4.20.

As can be seen, multiplying the adjacent bands in a nondecimated wavelet transform reduces the magnitude of coefficients that correspond to noise. This is because with the increase in scale, their magnitude decreases.

Similarly, the magnitude of the coefficients that correspond to singularities increases as, with the increase in scale, their magnitude increases. Applying the thresholding technique on these noise-suppressed bands will give us results that exploit the interscale dependencies. Bao and Zhang also suggested an adaptive threshold on the products rather than on the wavelet coefficients.

In this way, the wavelet transform helps in singularity detection and leads to the techniques of signal denoising. Now, we will see how a singularity detection method can be put to use in a real-world application. In the next section, we will discuss 1-D singularity detection in ECG signals and then 2-D singularities in proteomic images.

Two Case Studies of Singularity Detection

One-Dimensional Singularity in ECG Signals

The ECG is a graphic record of the direction and magnitude of the electrical activity [19] that is generated by depolarization and repolarization of the atria and ventricles. The ECG signal shown in Figure 4.21 carries a time-varying structure called the P-QRS-T complex. Most of the clinically useful information in the ECG is found in the intervals and amplitudes defined by its features (characteristic wave peaks and time durations, as shown in a sample in Figure 4.22). Signal frequencies are distributed as (1) low-frequency P and T waves

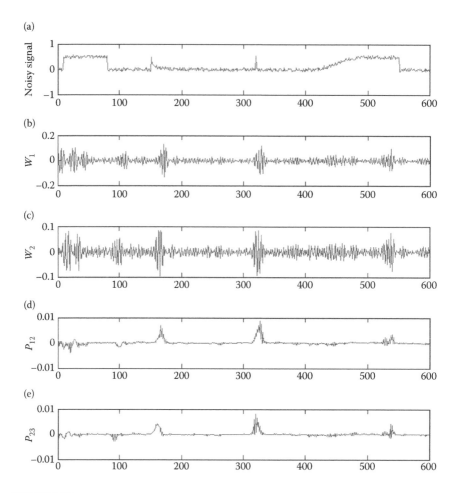

FIGURE 4.20
(a) Signal with additive random noise, (b) W_1: detail coefficients at scale 1, (c) W_2: detail coefficients at scale 2, (d) P_{12}: product of detail coefficients at scale 1 and 2, and (e) P_{23}: product of detail coefficients at scales 2 and 3.

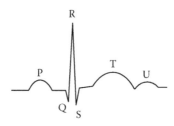

FIGURE 4.21
A typical cycle of ECG, with all the singularities shown.

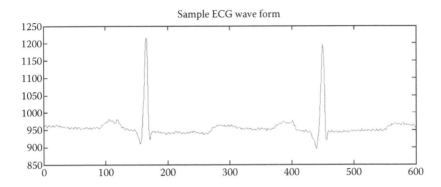

FIGURE 4.22
Normal sample of an ECG waveform.

and (2) mid- to high-frequency QRS waves. These features could be viewed as 1-D singularities. A system analyzing the ECG wave should, first and foremost, be accurate and reliable in detecting the QRS complex along with the P and T waves. P waves represent the activation of the atria, whereas the QRS and T waves represent the activation of ventricles. For taking the wavelet transform, the Daubechies 10 (DB10) wavelet is most suitable because its shape is the most similar to the ECG waveform compared with all other wavelets.

Mohmoodabadi et al. [19] evaluated an ECG feature extraction system. Because the peaks in the ECG signal have different frequencies, they utilized the unique property of decomposing the signal in different frequency bands of the wavelet transform to extract the positions of the peaks. They found that the wavelet filter with scaling functions that were more closely similar to ECG signals achieved better detection.

After applying the wavelet transform up to eight levels of decomposition, as shown in Figure 4.23, it can be observed that most of the details are contained at the D5 level. Also, it can be said that the levels D1 and D2 essentially capture the high-frequency noise part. High-frequency components decrease as we increase the scale, because lower details are removed from the original signal. With the increase in scale, the QRS peaks flatten and P and T waves are more visible due to their low-frequency content as discussed previously.

Figure 4.24 shows the frequency spectrum of the detail bands D1 to D8. As can be seen, information content in D1 and D2 is mainly above 45 Hz, which corresponds to noise. This is the reason that we will remove these detail coefficients while extracting the position of the peaks P-QRS-T.

After applying the wavelet transform, different algorithms are applied to detect different features in the ECG waveform. At first, detection of the R peak is the most important, as it defines the cardiac beats and accuracy of detection. All other features are dependent on it. The R peaks belong to the mid frequency region, so the details from 23 to 25 Hz are retained and the rest are made to zero. The reconstructed signal gives high amplitude at the R peak locations.

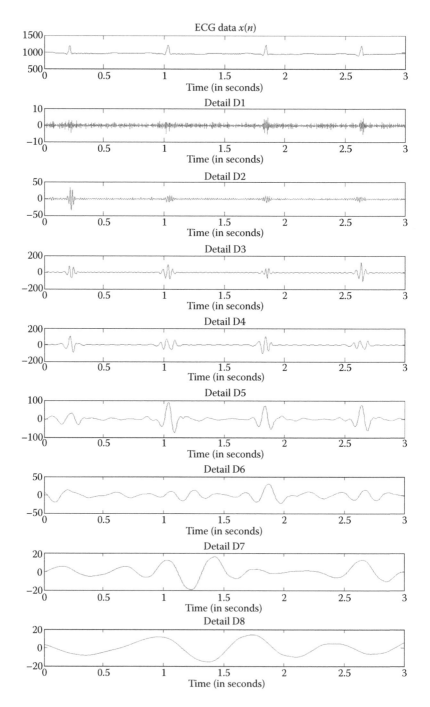

FIGURE 4.23
D1 to D8 details of an ECG sample.

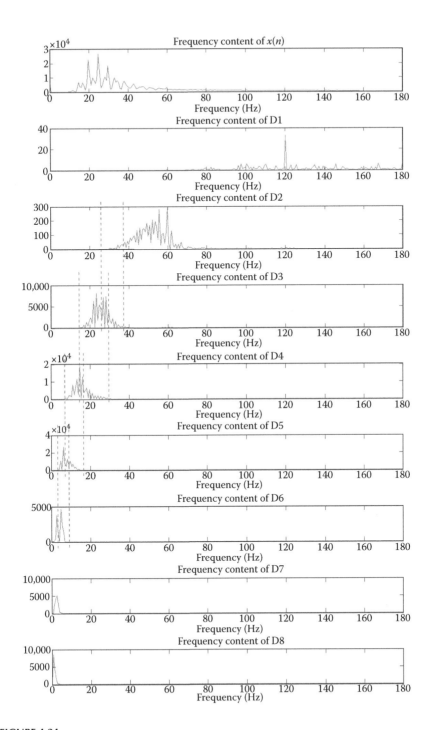

FIGURE 4.24
Frequency spectrums of D1 to D8 detail coefficients.

Q and S peaks occur on the side of the R peak. For Q and S detection, details up to 25 Hz are removed, leaving only the approximate signals. Now, we check for the peaks in the reconstructed signal in the neighborhood of the R peak positions. The left peak denotes Q and the right one denotes S. Similarly, for the P and T peaks, we retain details from 24 to 28 Hz and remove all other frequencies.

Two-Dimensional Singularities in Proteomic Images

Proteomics is the field that analyzes the structures, functions, and interactions of the proteins produced by the genes of a particular cell or tissue. The most important task in proteomics is to separate the proteins from the cell or the tissue. Two-dimensional gel electrophoresis is one of the most popular techniques for analyzing and separation of complex protein mixtures. In this technique, individual proteins are resolved in the first dimension according to their isoelectric point, and in the second dimension according to their molecular weight. This provides sufficient information about a variety of proteins simultaneously. This technique also has some drawbacks. The system nonlinearities involved in the gel formation process create a lot of nonlinear intensity variation in protein spots as well as in the background. Inevitably though, faint, saturated, and overlapped protein spots and streaks will appear. Therefore, these types of 2-D gel images constitute a good example of 2-D singularities.

Such large and unstructured variations make it impossible to distinguish low abundant protein spots from the background in the space or frequency domain alone. Figure 4.28 shows a typical gel image. It can be observed that most of the protein spots are closely either circular or elliptical in shape. The main challenge in segmenting these protein spots is in identifying their boundaries. The boundaries of these protein spots are not just sharp discontinuous points, but are typically located along smooth curves called *contours*. Separable wavelets are only good at isolating the sharp discontinuous points, but fail to identify the smoothness along these contours. Also, the separable wavelet transform captures the singularities in limited directions (horizontal, vertical, and diagonal). In general, the images are not a cross-product of two 1-D signals. Most of the images are inherently 2-D. Therefore, we need a tool that is capable of handling all these problems effectively; for example, 2-D nonseparable wavelets, which have been found to be the more effective option. The nonseparable wavelet is able to represent directional information in a better way, so it seems to be a promising tool for processing the 2-D gel images. Figure 4.25 shows the difference between wavelet transforms for a synthetic gel image using separable and nonseparable wavelets. It can be observed that the details obtained in the nonseparable wavelet domain are more significant compared with those obtained in the separable wavelet domain. Also, the edges are more sharply visible in the nonseparable domain.

In the context of wavelets and filter banks, nonseparable systems have nonrectangular frequency supports for the subbands, thus resulting in

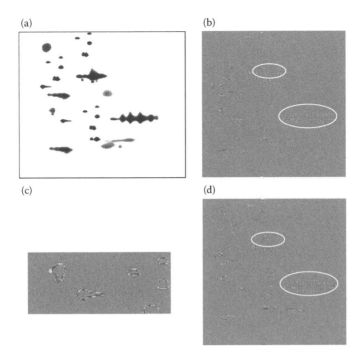

FIGURE 4.25
(a) A synthetic gel image, (b) wavelet transform obtained using separable wavelet transform, (c) wavelet transform using nonseparable wavelet transform, and (d) detail sub-band of quincunx wavelet.

better frequency selectivity as commonly required in two dimensions. The two-channel quincunx filter banks have received particular attention. In two dimensions, the 2-D quincunx lattice is the only nonseparable lattice that gives a subsampling factor of 2. The subsampling matrix $Q = \begin{bmatrix} 1 & 1 \\ 1 & -1 \end{bmatrix}$ generates the quincunx lattice.

The two-channel quincunx filter bank is shown in Figure 4.26. The quincunx low-pass and high-pass filters are often chosen to have diamond-shaped frequency supports, as shown in Figure 4.27. With these diamond-shaped frequency responses, the low-pass filter can preserve the high frequencies in the vertical and horizontal directions, which is a good match for the human visual system because the visual sensitivity is higher to changes in these two directions than in other directions. Due to this, quincunx filter banks are particularly important in image processing applications.

It is usually assumed that the wavelet coefficients representing singularities due to noise are usually spread out and have a low value, whereas singularities due to image features are sparse and produce strong wavelet coefficients. Based on this assumption, images are denoised in the wavelet domain without causing important details to deteriorate.

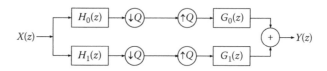

FIGURE 4.26
Quincunx filter bank.

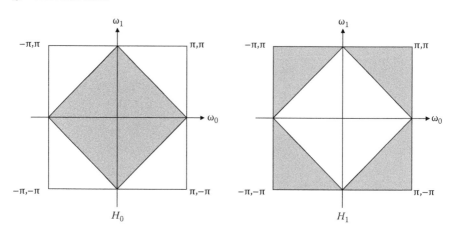

FIGURE 4.27
Quincunx filter bank frequency support.

After denoising, segmentation is another important part of the processing of the gel image, which extracts individual spots. Although denoising in the nonseparable domain gives the best result, a large number of nonlinear variations still pose problems for segmentation. The "watershed" is a popular medical image segmentation method. The available segmentation methods, including the watershed, use various thresholds for the identification of the regions where protein spots are present. A lower value of the threshold results in several artifacts being detected as spots whereas a higher threshold value does not help in detecting many spots. Depending on the nature of the gel images, global thresholds (i.e., a common threshold for distinguishing different regions of a gel image) cannot work and local thresholds used for identifying the different regions of the gel image are hard to find. For the purpose of illustration, in the concept of interscale dependencies introduced by Bao and Zhang [18], a protein spot is extracted manually from the proteomic image and is used as our reference image in Figure 4.28.

We have studied singularity detection in 2-D gel images across scales in the nonseparable domain and it has been found that it can provide valuable hints for image segmentation. A gel image is first decomposed using the undecimated quincunx wavelet transform to maintain the shift invariance of the transformed image. In the undecimated transform, decimation operators are removed and filters are upsampled at each of the next scales. It has

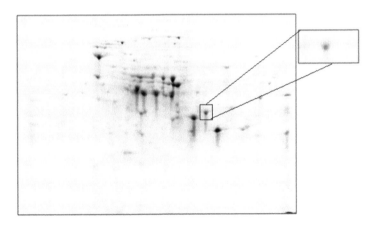

FIGURE 4.28
Left, a typical gel image; right, spot extracted for processing.

been observed that singularities caused by background and noises can be decorrelated across scales up to some extent. Therefore, a scale product (*P*) between two adjacent scales (i.e., *j* and *j* + 1) is defined as an element-wise multiplication of the detail signal at scale *j* and the detail signal at scale *j* + 1.

On the extracted spot, we applied a two-level undecimated quincunx wavelet transform. To suppress the noise and enhance the coefficients corresponding to singularities, a scalar product of the two high-frequency bands obtained is defined.

Figure 4.29 shows the surface plot of the original spot segmented, with the amplitude of the spot at approximately 140 and the noise at level 20. On the

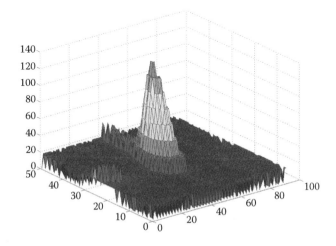

FIGURE 4.29
Parametric surface of the original spot.

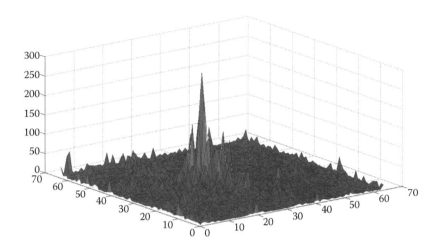

FIGURE 4.30
Parametric surface of the subband obtained after adding the product of adjacent subbands.

other hand, Figure 4.30 shows the scalar product of the detail bands with the amplitude of the coefficient around the spot at approximately 350 and the noise remaining the same at level 20. In Figure 4.30, peaks essentially represent the boundary points of the spots.

As mentioned above, for processing the actual proteomic image, watershed segmentation is applied to divide the image into homogenous parts for further processing. To prevent oversegmentation, watershed segmentation is applied on the low-resolution version of the image. Resultant partitions obtained from watershed segmentation are mapped onto the scale product. This strategy enables us to study singularities within a local area (or partition) of a scale product, which in turn helps in detecting the boundaries of the spots. These boundaries are then connected using a connected maxima set method, which is described in our previous work [20]. Figure 4.31 shows

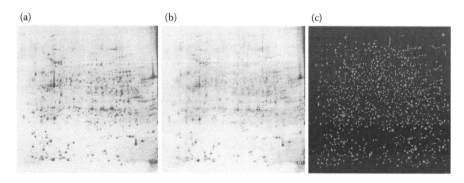

FIGURE 4.31
(a) Original proteomic image, (b) segmented image, and (c) connected maxima set.

singularity detection and the boundaries extracted in proteomic images using this method.

To conclude, we have presented several examples to explain how the wavelet transform is useful in data compression—from the perspective of self-similar properties, localization to one scale and persistence across scales, singularity detection, and signal denoising. There are several other ways in which interscale dependencies in the wavelet transform have been exploited to advantage. However, for brevity, we shall conclude the chapter here.

Acknowledgments

The authors gratefully acknowledge the Board for Research in Nuclear Sciences, BARC, Mumbai for supporting this research work and for the material on proteomics discussed at various points in the chapter. We also collectively acknowledge the assistance received from Rohanil Raje and Satyaprakash Pareek, who were students of the Dual Degree (B. Tech. + M. Tech.) Programme in the Department of Electrical Engineering, Indian Institute of Technology Bombay. In developing the contents of the chapter, we have often taken standard "benchmark" images like "Lena" in Figures 4.4 and 4.19, and in a typical 2-D fractal in Figure 4.14. We place on record, our gratitude to the several people who have made these images available for standard tests in the literature. We have taken the liberty to use these standard images and 2-D data freely, assuming that all people who have contributed to making these a standard in the field would concur with our view that these are, indeed, benchmarks that can thus be freely used. We also thank Dr. Aditya Abhyankar, Professor, University of Pune, for his very constructive reviews of Chapters 2 and 4 of this book, which helped us reformulate these chapters in the form presented here. Although we have tried our best to acknowledge the contributions of all those who have helped the authors create these two chapters, we tender an apology here to anyone whom we have, inadvertently, omitted from explicit acknowledgement, with an assurance that we are, indeed, grateful for the assistance that they have rendered.

References

1. Said, A., Pearlman, W. "A new, fast and efficient image codec based on set partitioning in hierarchical trees," *IEEE Transactions on Circuits and Systems for Video Technology*, vol. 6, pp. 243–250, 1996.

2. Shapiro, J.M. "Embedded image coding using zerotrees of wavelet coefficients," *IEEE Transactions on Signal Processing*, vol. 41, no. 12, pp. 3445–3462, 1993.

3. Gadre, V.M. "Video lectures: Advanced digital signal processing-multirate and wavelets," Available: http://nptel.iitm.ac.in/courses/117101001/28.

4. Peter R. Massopust, *Fractal Functions, Fractal Surfaces, and Wavelets*. Academic Press, San Diego, CA, 1994.

5. Mallat, S. *A Wavelet Tour of Signal Processing*. Academic Press, San Diego, CA, 1998.

6. Wornell, G.W., Oppenheim, A.V. "Wavelet based representations for a class of self-similar signals with application to fractal modulation," *IEEE Transactions on Information Theory*, vol. 2, pp. 785–800, 1992.

7. Miduri, S., Pagano, M., Russo, F., Tartarelli, S., Giordano, S. "A wavelet-based approach to the estimation of the hurst parameter for self-similar data," *Proc. 13th Int. Conf. Digital Signal Processing (DSP '97)*, vol. 2, pp. 479–482, 1997.

8. Flandrin, P. "Wavelet analysis and synthesis of fractional Brownian motion," *IEEE Transactions on Information Theory*, vol. 38, no. 2, pp. 910–917, 1992.

9. Meyer, Y. *Ondelettes et Operateurs*, vol. I–III. Hermann, Paris, 1990.

10. Mallat, S., Hwang, W.L. "Singularity detection and processing with wavelets," *IEEE Transactions on Information Theory*, vol. 38, no. 2, pp. 617–643, 1992.

11. Pizurica, A., Philips, W., Lemahier, I., Acheroy, M. "A versatile wavelet domain noise filteration technique for medical imaging," *IEEE Transactions on Medical Imaging*, vol. 22, no. 3, pp. 323–331, 2003.

12. Donoho, D.L., Johnstone, I.M. "Adapting to unknown smoothness via wavelet shrinkage," *Journal of the American Statistical Association*, vol. 90, no. 432, pp. 1200–1224, 1995.

13. Chang, S.G., Bin, Y., Vetterli, M. "Adaptive wavelet thresholding for image denoising and compression," *IEEE Transactions on Image Processing*, vol. 9, no. 9, pp. 1532–1546, 2000.

14. Johnstone, I.M., Donoho, D.L. "Ideal spatial adaptation via wavelet shrinkage," *Biometrika*, vol. 81, no. 3, pp. 425–455, 1994.

15. Xu, Y., Weaver, J.B., Healy, D.M., Jian, L. "Wavelet transform domain filters: A spatially selective noise filtration technique," *IEEE Transactions on Image Processing*, vol. 3, no. 6, pp. 747–758, 1994.

16. Sadler, B.M., Swami, A. "Analysis of multiscale products for step detection and estimation," *IEEE Transactions on Information Theory*, vol. 45, no. 3, pp. 1043–1051, 1999.

17. Mallat, S., Zhong, S. "Characterization of signals from multiscale edges," *IEEE Transactions on, Pattern Analysis and Machine Intelligence*, vol. 14, no. 7, pp. 710–732, 1992.

18. Bao, P., Zhang, L. "Noise reduction for magnetic resonance images via adaptive multiscale products thresholding," *IEEE Transactions on Medical Imaging*, vol. 22, no. 9, pp. 1089–1099, 2003.

19. Mohmoodabadi, S.Z., Ahmadian, A., Abolhasani, M.D. "ECG feature extraction using daubechies wavelets," *Proc. of the Fifth IASTED International Conference*, Benidorm, Spain, 2005.

20. Sengar, R.S., Upadhyay, A.K., Singh, M., Gadre, V.M. "Segmentation of two dimensional electrophoresis gel image using the wavelet transform and the watershed transform," *2012 National Conference on Communications (NCC)*, pp. 1–5, 3–5, 2012.

5

Fractals and Wavelets in Applied Geophysics with Some Examples

R. P. Srivastava

CONTENTS

Introduction 155
Fractal Time Series Characterization 156
    Concept of fBm and fGn 156
R/S Analysis 157
Determination of Fractal Dimension: Box Counting Method 159
    Box Counting Method 159
Design of Survey Network 162
    Spectral Approach 162
    Fractal Dimension Approach 163
    Optimum Grid 164
    Detectability Limit of the Survey 166
Fast Multiplication of Large Matrices: A Useful Application for
Gradient-Based Inversion 167
Wavelet Transform of Gravity Data for Source Depth Estimation 170
Acknowledgments 173
References 173

Introduction

Wavelets and fractals are relatively new tools in geophysical analysis; however, in the last two decades, they have gained substantial popularity due to their applications in nonlinear signal analysis (Dimri 2000; Dimri et al. 2012). The essence of fractal analysis lies in fractal dimension analysis. Furthermore, the concept of fractal analysis is extended to time series analysis using the concept of fractional Brownian motion (fBm) and fractional Gaussian noise (fGn). A widely used concept called the box counting method for fractal dimension analysis is explained with examples. Wavelet analysis can be seen as an advanced substitute for the Fourier analysis, which has been widely used in geophysics for noise attenuation, power spectrum analysis (Maus and Dimri 1995a,b), characterization of fractal/scaling behavior of the time/space series using power spectrum approach (Dimri et al. 2011; Srivastava et al. 2007; Dimri and Vedanti 2005) and frequency analysis of the seismic time series (Dimri 1992; Srivastava and Sen 2009, 2010), and many more similar applications. An intriguing application of wavelet transform is given in Moreau et al. (1999), in which they present the estimation of source parameters, that is, horizontal location, depth, multipolar nature, and strength of the homogeneous potential field sources based on a special class of wavelets that are invariant while upward continuation. This makes source parameter analysis very easy and intuitive simply by looking at the wavelet transform, whereas quantitative estimates can be drawn with the help of lines of extrema of the wavelet transform. Other applications of wavelet-based analysis in geophysics include inversion of geophysical data using wavelets (Li et al. 1996) and fast computation of the inverse of large matrices. In this article, some examples of wavelet transform for fast computation of large matrix inversion and source parameter estimation from gravity data are presented.

Fractal Time Series Characterization

Concept of fBm and fGn

The fractal analysis of time series is fundamentally based on the concept of fBm and fGn. The concept of fBm and fGn has subtle differences and are mathematically intertwined (Hewett 1986; Hardy and Beier 1994). To explain it with a sequence of numerical observations, for instance, if a time series with a time interval of Δt representing the fBm sequence is given by $t_1, t_2, t_3, \ldots t_{n-1}, t_n$, then a time derivative sequence of fBm is known as fGn, which can be generated as $[(t_2 - t_1), (t_3 - t_2), \ldots (t_n - t_{n-1})]/\Delta t$. Conversely, if a sequence of numbers with time interval Δt representing fGn is given by $[g_1, g_2, g_3, \ldots g_{n-1}, g_n]$, then a corresponding fBm can be computed by simple integration given by $[g_1 + g_2 + g_3 \ldots g_{n-1} + g_n] \Delta t$, or in terms of sequence, it will be written as $[g_1, g_1 + g_2, g_1 + g_2 + g_3, \ldots]$.

To test for a fractal characteristic of fBm sequence, the following definition of fBm can be used.

fBm is a single valued continuous function of time, that is, $f(t)$, having the following properties:

1. The increments, i.e., $f(t_2) - f(t_1)$ follow Gaussian distribution
2. $E\{f(t_2) - f(t_1)\} = 0$, i.e., expectation of increments is zero or close to zero
3. $E\{[f(t_2) - f(t_1)]^2\} = \alpha |t_2 - t_1|^{2H}$, i.e., expectation of increment square follows power law

where E is the expectation, α is the constant, and H is the Hurst exponent.

The fractal time series, also known as a time series having long memory, is characterized by fGn. A time series is said to follow fGn characteristics (*i*) if its histogram is fat-tailed Gaussian, (*ii*) covariance, semivariogram, and power spectrum follows power law behavior, and (*iii*) Hurst coefficient lies within (0.5–1.0). It is important to note that according to Hewett (1986), many observations suggest that in sedimentary environments, vertical observation of values such as well logs show fGn characteristic whereas lateral variations show fBm characteristics.

In the above stated tests, the histogram is a simple statistical measure, and if the histogram follows a long-tailed Gaussian distribution or Levy distribution, then we assume it to display fGn behavior. Furthermore, the covariance (Mandelbrot and van Ness 1968; Srivastava and Sen 2009) and variogram (Hardy and Beier 1994) of fGn is given by following Equations 5.1 and 5.2, respectively:

$$\text{Cov}(\tau) = \frac{1}{2}\left(|\tau - 1|^{2H} - 2|\tau|^{2H} + |\tau + 1|^{2H} \right) \tag{5.1}$$

$$\gamma(\tau) = a - b|\tau|^{2H-2} \tag{5.2}$$

where τ is time lag and H is Hurst coefficient.

Another important test is power spectrum evaluation, which is explained in detail in Chapter 1 of this book.

Furthermore, to confirm the fGn behavior of a time series, it is important to compute the Hurst coefficient, which characterizes the time series. For instance, $H = 0.5$ corresponds to random noise, and $0.5 < H \leq 1.0$ signifies stronger and stronger long-range dependence in the time series. For more details, readers are encouraged to read Dmowska and Saltzman (1999). There are various methods to compute Hurst coefficient (Caccia et al. 1997); however, the simplest and most popular method is R/S analysis. A brief description of R/S analysis is given below.

R/S Analysis

It is known that if the input signal has a sufficient number of samples, then a rescaled range (R/S) analysis works well. Hurst et al. (1965) found empirically that many data sets in nature satisfy the power law relation:

$$\frac{R}{S} = \left(\frac{N}{2} \right)^{H}, \tag{5.3}$$

where H is the Hurst coefficient, N is the number of data points, and R and S stand for the range and standard deviation obtained from the R/S analysis.

The R/S analysis is easily extended to discrete time series. The analysis is based on the computation of range and standard deviation of the detrended time series. The running sum of the series relative to its mean is computed for several segments of the time series, which is given by

$$y_n = \sum_{i=1}^{n} (y_i - \bar{y}_n), \quad n = 2, N-1 \tag{5.4}$$

The range is defined by

$$R_N = (y_n)_{max} - (y_n)_{min} \text{ with } S_N = \sigma_N \tag{5.5}$$

where S_N is standard deviation of the each segment of the time series (y_n).

The plot of $\log(R/S)$ versus $\log(N/2)$ is a straight line, and the slope gives value of the Hurst coefficient H.

Now, with the above background, given a time series, we are ready to test its behavior for fGn characteristics, and if it shows fGn characteristics, then it implies that time series has long-range dependence on its previous values and is known as a fractal time series. A porosity log is analyzed for its fractal behavior using the criteria shown in Figure 5.1. Note that the porosity

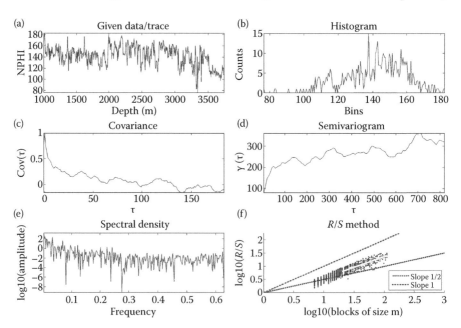

FIGURE 5.1
Test of fractal behavior of the time series: (a) porosity log, (b) histogram of the well log, (c) covariance showing power law behavior, (d) semivariogram showing power law behavior, (e) log power spectrum showing power law, and (f) Hurst coefficient ($H = 0.81$) showing fGn characteristic of the well log.

log follows fractal behavior, and thus it provides an opportunity to generate synthetic logs in those areas in which wells are not available using the fractal characteristic of available wells assuming similar geological settings.

Determination of Fractal Dimension: Box Counting Method

There are several applications of fractal dimension; however, the most prominent application is to determine length of curved line, the area of an irregular two-dimensional (2-D) object (e.g., a leaf) and the volume of an irregularly shaped three-dimensional object (e.g., a rock mass). Some applications of fractal dimensions in geophysics include fractal dimension analysis of earthquakes (Dimri et al. 2005), planning of survey network (for instance, a gravity and magnetic survey network) keeping in view the detectability limit of the survey (Dimri 1998; Srivastava et al. 2007).

Box Counting Method

The box-counting method is considered to be the most robust method for the estimation of fractal dimension because it can be applied with equal effectiveness to point sets, linear features, areas, and volumes. The following steps are employed in the process of estimation of fractal dimension using the box counting method:

1. Cover the given feature with a single "box"
2. Divide the box into four quadrants and count the number of cells that occupy some portion of the given feature (say a curved line, or spatial data locations, etc.)
3. Divide each subsequent quadrant into four subquadrants, and continue to do so until the minimum box size is equal to the resolution of the data

Mathematically, if n = number of boxes filled with the feature of which fractal dimension is being determined, b = box size, then it is observed that n, b follows the relation:

$$n \propto b^{-D} \tag{5.6}$$

where D is the fractal dimension. The plot between $\log(n)$ and $\log(b)$ gives negative slope that gives the value of the fractal dimension of the given feature. The above process of fractal dimension determination is shown in the following figures. Figure 5.1 shows a curved line whose fractal dimension is to be determined. Also, it is shown that the curved line is encompassed in a square. Furthermore, in Figure 5.2a, it is shown that the square encompassing the curved line is further divided into for equal-sized squares, and so on, as shown in Figure 5.2b. The finer the division, the better the estimate of the fractal dimension; however, there is a limit to dividing the square as discussed in Srivastava et al. (2007) and Keating (1993). Finally, Figure 5.3 shows a log-log plot of the box size versus the corresponding number of only those cells (boxes) containing some part of the feature whose fractal dimension is being determined. The slope of the log-log plot gives the fractal dimension of the object.

Similarly, an example for computing the fractal dimension of the point data set is given in Figure 5.4. Theoretically, for the point data set, boxes

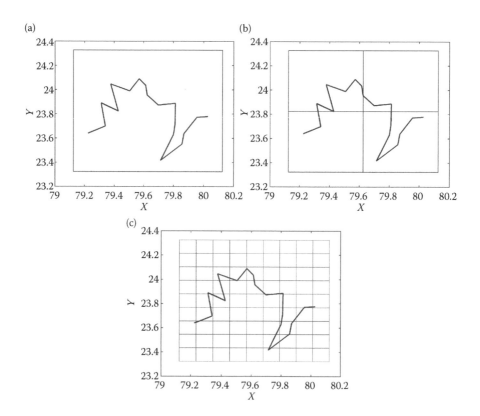

FIGURE 5.2
(a) Curved line object for which fractal dimension is to be determined. The first step is to enclose the curved line within a square, as shown above with a dark line. (b) As shown in (a), the single square is further divided into four equal parts. As explained above, in this case, all four squares are counted because they contain some parts of the curved line whose fractal dimension is to be determined. (c) As shown in (b), further squares are subdivided, and only those squares that contain some part of the curved line are counted.

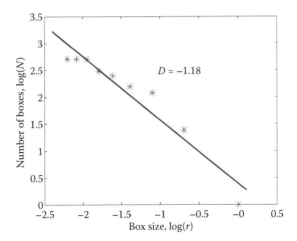

FIGURE 5.3
Log-log plot of box size versus corresponding number of boxes containing some part of the feature whose fractal dimension is being determined. The slope of the best fit line provides the fractal dimension.

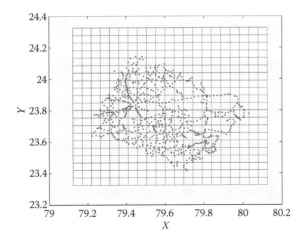

FIGURE 5.4
Point data set, as an input to fractal dimension analysis using the box counting method.

should be divided until each box contains only a single data point. Once each box contains only a single data point, then further division will not increase the number of boxes because only those boxes that contain some part of the observation data are considered in fractal dimension analysis. Thus, Figure 5.4 still needs further subdivision of the boxes for more accurate fractal dimension determination. Figure 5.5 shows the log-log plot of the box size and number of boxes, and the slope gives approximate fractal dimension of the point data distribution shown in Figure 5.4.

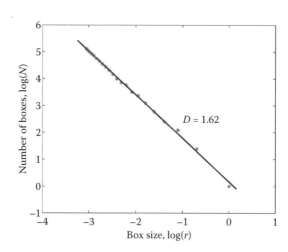

FIGURE 5.5
Log-log plot of box size versus corresponding number of boxes containing data points. The slope of the best fit line gives the fractal dimension of the point data distribution.

Design of Survey Network

Often, gravity/magnetic data for regional studies is acquired along accessible roads/tracks and pedestrian ways, and then the results are presented as contour maps or color maps employing that the survey grid defines the observed field continuously to justify interpolation. Normally, such surveys fulfill the need for drawing a profile across a known geological structure, but this method of acquisition is not acceptable for 2-D surveys. It is seen that the data acquired along roads and tracks generates lots of spurious anomalies due to aliasing after interpolation. To avoid such problems, two strict measures should be taken: (1) the data should be acquired at regular intervals. Normally, this is not possible due to logistic problems, in which case, care should be taken to acquire the data in such a way that it fills the area of study to maximum extent; and (2) while gridding the data for further processing and interpretation, an optimum gridding interval should be chosen to avoid aliasing caused by the interpolation of the data.

A survey grid (network) can be designed using spectral theory (Reid 1980) and fractal concepts. Synthetic survey grids are optimized using fractal dimension analysis keeping in view the optimum number of stations and the detectability limit of the survey.

Spectral Approach

Let Δx be the station interval (or in the case of 2-D data grid interval), and λ_N be the shortest wavelength that can be correctly identified by this grid, then the Nyquist wavelength is given by $\lambda_N = 2\ \Delta x$. Thus, any wavelength shorter than λ_N will be aliased into a wavelength longer than λ_N (Blackman and Tukey 1958). According to Shannon's sampling theory for a 2-D case, the number of samples (n) necessary to restore the field is given by (Brillouin 1962)

$$n = \pi x^2 / \lambda_N, \tag{5.7}$$

where x is the size of the square region/survey area and λ_N is the shortest wavelength.

Now, if we interpret the same concept in terms of power spectrum, then it should not contain significant power at wavelengths shorter than the chosen λ_N. It is thus important to decide the Δx and design the survey network depending on the needs of the investigation, keeping in view the depth of investigation (z). The fraction of power aliased can be determined following Reid (1980). Mathematically, it can be shown that the expectation value of the power spectrum $\langle E(r) \rangle$ for an assemblage of small sources of considerable depth extent, is given by (Spector and Grant 1970)

$$\langle E(r) \rangle = 4\pi^2 k^2 \exp(-2zr) \tag{5.8}$$

where z is the elevation difference between the top surface of the magnetic/gravity block and the sensor, k is the mean value of the magnetic moment per unit depth, and r is the wave number. For a given survey spacing Δx, there will be a Nyquist wave number given by, $w_N = 2\pi / \lambda_N$, then the aliased fraction of the power (F) may be determined as

$$F = \left[\int_{r_N}^{\infty} \langle E(r) \rangle dr \right] \div \left[\int_0^{\infty} \langle E(r) \rangle dr \right] = \exp(-2\pi z / \Delta x) \tag{5.9}$$

Fractal Dimension Approach

In fractal dimension analysis of spatial data (Turcotte 1997), the fractal dimension of the survey network (location of data points) is calculated. This tells about the coverage of the area. High fractal dimensions imply better coverage and vice versa. It is demonstrated in the following figures that homogeneous distribution of data points results in higher fractal dimension compared with the same number of data points distributed in a clustered manner as shown in Figures 5.6 and 5.7, respectively. Thus, it is recommended that the acquired data be homogeneously distributed rather than being clustered in a particular region due to ease of access.

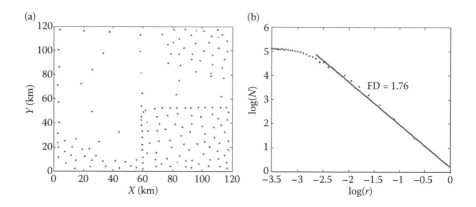

FIGURE 5.6
(a) Inhomogeneous data distribution (number of points = 166). (b) Fractal dimension of inhomogeneous data distribution.

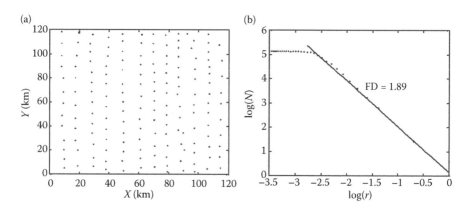

FIGURE 5.7
(a) Homogeneous data distribution (number of points = 166). (b) Fractal dimension of homogeneous data distribution.

Optimum Grid

Several synthetic grids for gravity survey design, were generated in the area of study. However, selection of optimum grid to acquire the data depends on several governing factors, in which a few of them include:

1. *Accessibility of the Terrain.* Based on this information, select the grid spacing to acquire the data at regular intervals. If the terrain is difficult, then long spacing, which can be achieved while surveying, may be selected.

2. *Resolution of the Survey.* The spacing between any two consecutive stations defines the horizontal resolution of the survey. This is another parameter that plays an important role in survey grid selection, but this parameter depends on other parameters such as the size of the target body given in following point no. 3.

3. *Size of the Target Body.* There is a trade-off between the size of the body under investigation and grid spacing. For the purpose of petroleum exploration, the size of the area is often very large, and for mineral exploration, smaller areas are surveyed in detail. In Figure 5.8, the grids at 2, 3, 4, and 5 km grid intervals are analyzed using fractal dimension as shown in Figure 5.9. Finally, the results can be tabulated to have a comprehensive look and then an optimum grid can be selected based on the above points 1 to 3.

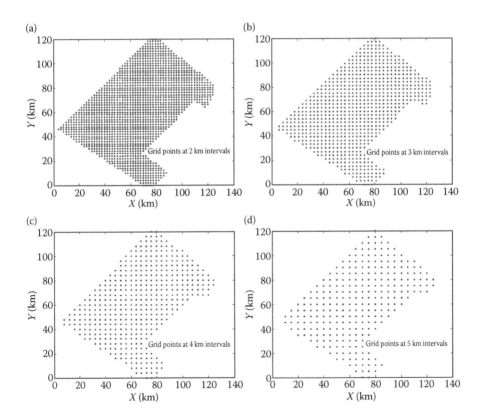

FIGURE 5.8
Synthetic grid at (a) 2 km, (b) 3 km, (c) 4 km, and (d) 5 km intervals.

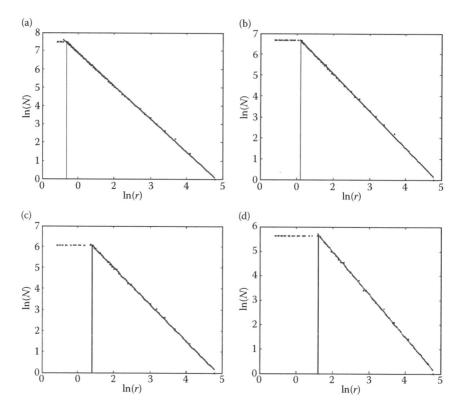

FIGURE 5.9
Fractal dimension of synthetic grid points at (a) 2 km spacing (FD = 1.79), ln(*r*) = 0.71, *r* = 2.03 km; (b) at 3 km spacing (FD = 1.76), ln(*r*) = 1.12, *r* = 3.06 km; (c) at 4 km spacing (FD = 1.74), ln(*r*) = 1.40, *r* = 4.05 km; and (d) at 5 km spacing (FD = 1.73), ln(*r*) = 1.62, *r* = 5.05 km.

Detectability Limit of the Survey

The detectability limit of any geophysical survey network (i.e., station spacing in case of gravity and magnetic survey) depends on the fractal dimension of the survey network and the anomaly (Lovejoy et al. 1986; Korvin et al. 1990; Korvin 1992; Dimri 1998). Geophysical anomalies due to fractal sources such as density and susceptibility distribution (Pilkington and Todoeschuk 1993; Pilkington et al. 1994; Maus and Dimri 1994, 1995a,b, 1996) cannot be accurately measured if its fractal dimension does not exceed the difference of the 2-D Euclidean and fractal dimension of the network (Lovejoy et al. 1986; Korvin 1992). Mathematically, the detectability limit of geophysical surveys is given by Korvin (1992) as

$$Ds = E - Dn \qquad (5.10)$$

TABLE 5.1

Estimation of Fractal Dimension, Detectability Limit with Number of Station, and Corresponding Station Spacing of 2, 3, 4, and 5 km in a Study Area of 8,035.78 km^2

Station Spacing (km)	Fractal Dimension of Designed Grid	No. of Stations	Detectability Limit
2	1.79	2005	0.21
3	1.76	886	0.24
4	1.74	504	0.26
5	1.73	325	0.27

where Dn and Ds are the fractal dimensions of the network and the source, respectively, and E is the Euclidean dimension (which is 2 in case of 2-D survey). The detectability limits for a survey designed with station spacing at 2, 3, 4, and 5 km is shown in Table 5.1. The table shows detectability limits versus no. of stations (data points) for different station spacings to complete the survey. The aim is to optimize higher detectability with lower cost by choosing the appropriate station spacing.

It is well known that if the fractal dimension (FD) of the measuring network is 2, then the network can detect objects of any dimension (Lovejoy et al. 1986; Dimri 1998). This reveals that for best detectability, one should try to design a measuring network with a dimension of approximately 2. It was also observed in earlier studies that if FD is much less than 2, the field being sampled cannot be fully recovered and some aliasing will occur (Lovejoy et al. 1986; Keating 1993). However, it is worth mentioning here that Equation 5.10 is valid only for the outcropping bodies because it has no relation with the depth values. Hence, the detectability limit of an object lying beneath the surface of the earth at depth z cannot be determined by the above equation. For such instances, an expression relating Mandelbrot fractal function and Fourier power spectrum (Berry and Lewis 1980; Higuchi 1990) can be used, which is given as

$$P(s) = s^{-\beta} = s^{-(5/2-D)} \tag{5.11}$$

where s is the wave number, D = fractal dimension of the source, and β is the scaling exponent.

Fast Multiplication of Large Matrices: A Useful Application for Gradient-Based Inversion

Matrix multiplication is an integral part of geophysical data analysis. In particular, those interested in the inversion of large data sets will find it more useful to get acquainted with the fast multiplication of the matrix and vector. In inverse theory, we usually solve a linear equation $d = Gm$ involving a matrix and vector multiplication (Dimri 1992; Sen and Stoffa 1995; Vedanti et al. 2005), where d is the data vector, G is the transformation matrix, and m is the model vector. The solution of the above equation for m also involves matrix–matrix and matrix–vector multiplication. Often, such multiplications are repeated several times based on the number of iterations the algorithm takes to converge to a desired solution. Hence, it becomes significantly important to minimize the multiplication cost. Wavelet-based matrix multiplication exploits the concept of wavelet decomposition. Because we know that wavelet decomposition reduces the size of the signal with level of decomposition, the higher the level of decomposition, the smaller the signal size and vice versa. It is this wavelet-based decomposition of the matrix that makes the matrix multiplication fast at the expense of accuracy. Although the accuracy of multiplication is compromised in this method, it is always within the permissible limits of error compared with conventional multiplication. The error increases if the matrix is very dense.

Wavelet-based multiplication is achieved in the following two steps:

Step 1: This is done only once, compute the matrix approximation, say (Am), at a suitable level k.

Step 2: This step is executed as many times as desired. Say, in case of geophysical inversion if the algorithm takes N iterations, then the following steps can be done N times, which is further subdivided into the following three steps: (a) compute vector approximation, (b) compute multiplication in wavelet domain, and (c) reconstruct vector approximation.

The execution time difference between conventional and wavelet-based multiplication becomes significant if the multiplication is repeated several times. However, if the multiplications is done only once or say for a few iterations (<20), then it is better to follow the conventional method of matrix multiplication.

Below is a sample MATLAB® code, which uses wavelet toolbox functions to achieve the matrix–vector multiplication. The time difference between conventional and wavelet-based multiplication versus iteration is plotted in Figure 5.10.

```
%% Fast multiplication of matrices
n = 1024*3;
lev = 5;
wav = 'db1';% name of the wavelet used 'Daubechies 1'
r = 100;%number of vectors
vec = zeros(r,n);
for i = 1:r
```

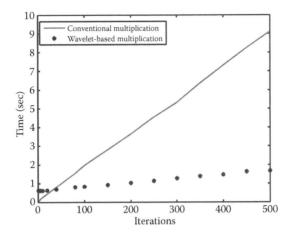

FIGURE 5.10

Comparison of the multiplication time taken by the wavelet method and conventional matrix multiplication method with a vector. Iteration shows number of times the multiplication of a fixed matrix with a vector, and time is the time taken by the above two methods. Continuous line shows time taken by the conventional method whereas dot (•) shows time taken by the wavelet method. It is remarkable to note that the wavelet-based method is very efficient as it takes much less time compared with the conventional method if the number of iterations is higher.

```
   vec(i,:)  =  (1:n);
end
vec = vec';
% Matrix is magic(1024*3)
m = magic(n);
[Lo_D,Hi_D,Lo_R,Hi_R]  = wfilters(wav);

% ordinary matrix multiplication by a vector.
tic
for num = 1:r
   p = m * vec(:,num);
end
toc
% Compute matrix approximation at level 5.

tic
sm = m;
for i = 1:lev
sm = dyaddown(conv2(sm,Lo_D),'c');
sm = dyaddown(conv2(sm,Lo_D'),'r');
end

% The three steps:
% 1. Compute vector approximation.
% 2. Compute multiplication in wavelet domain.
% 3. Reconstruct vector approximation.
```

```
for num = 1:r
sv = vec(:,num);
for i = 1:lev
  sv = dyaddown(conv(sv,Lo_D));
end
  sp = sm * sv;
for i = 1:lev
  sp = conv(dyadup(sp),Lo_R);
end
  sp = wkeep(sp,length(vec(:,1)));
end
toc
% Relative square norm error in percent when using wavelets.
rnrm = 100 * (norm(p-sp)/norm(p));
%rnrm
```

Wavelet Transform of Gravity Data for Source Depth Estimation

Wavelet-based depth estimation of the potential field sources is a very popular method (Moreau et al. 1997; Hornby et al. 1999). The wavelet-based method has another advantage in that before estimating the depth values, wavelet transform offers a very effective tool (wavelet denoising) to remove the noise from the data. Mathematical details and applications of the depth estimation using wavelet transform can be found in studies by Moreau et al. (1999) and Chamoli et al. (2006). Moreau et al. (1997) introduced a new class of wavelets called Poisson semigroup of wavelets, which follows Poisson semigroup properties. The kernel of Poisson semigroup of wavelets is given by

$$P_z(x) = \frac{1}{\pi} \frac{z}{z^2 + x^2} \tag{5.12}$$

Using derivatives of the above kernel, several wavelets can be generated and they belong to the Poisson semigroup of wavelets. The wavelet used in the following synthetic example is obtained by the first derivative $\left(\dfrac{\partial}{\partial x}\right)$ of Equation 5.12, assuming $z = 1$, which is given as follows:

$$P_1'(x) = \frac{-2x}{\pi(1 + x^2)^2} \tag{5.13}$$

where x could be assumed as a horizontal position, and z denotes scales or dilation of the wavelet. The plot of the wavelet given in Equation 5.13 is shown in Figure 5.11.

It is important to notice that the effect of wavelet transform of the gravity data using Equation 5.13 has the same effect as performing an upward continuation of the gravity data (potential field) by $z = 1$, and then taking a first derivative with respect to horizontal positions (x). Thus, the Poisson semigroup of wavelets is invariant with the upward continuation operator.

Wavelet transform has a nice geometrical property that the point where derivative of transforms with respect to translation (b) is zero, that is, $\dfrac{\partial^n}{\partial b^n} W(b,z) = 0$ (when $z \to 0$, where z is scale) gives the point of homogeneity center (location of the source). The lines which converge toward homogeneity center are called wavelet ridges of the wavelet transform and they form a cone whose apex coincides with the homogeneity center. However, this is true for homogeneous sources. In case of a potential field, the homogeneity center (source location) lies below the surface, hence the extrema lines of the wavelet transform intersects below $z = 0$ (scale = 0), and intersection of these extrema lines gives the source depth. There is a great deal of literature available on this method for a single source and multiple sources with inhomogeneity (Sailhac et al. 2000; Mauri et al. 2011).

Below is a synthetic example of gravity data analysis using the above wavelet to obtain the location and depth of the source. Synthetic gravity data over a sphere with a radius of 2 km located at a depth of 5 km below the surface having its center coordinate (5, 5) is obtained along a profile from 0 to 10 km at intervals of every 20 m. This is an example of a single homogeneous source for which wavelet transform using Poisson wavelet has been calculated. The wavelet transform shown in Figure 5.12a clearly shows the center of homogeneity at a horizontal location $x = 5$ km, which

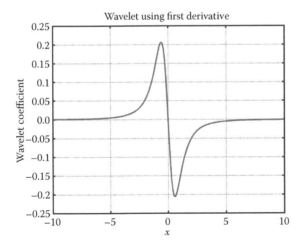

FIGURE 5.11
Wavelet using first derivative of the Poisson kernel (Equation 5.13).

is at the center of the sphere. The anomaly due to the sphere is also shown in Figure 5.12b. The lines passing through the minima and maxima of the wavelet coefficients are intersecting at $z = 5.2$ km, which gives the depth of the source.

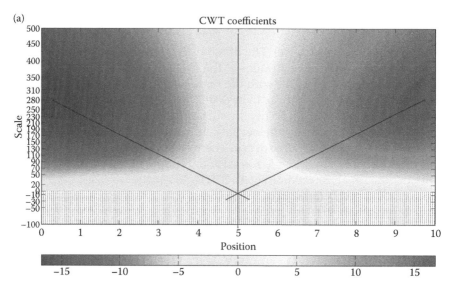

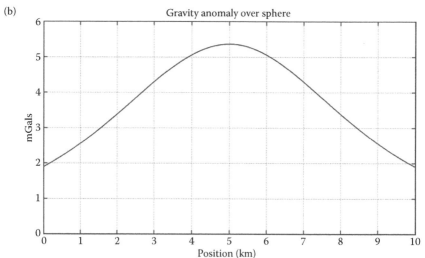

FIGURE 5.12
(a) Wavelet transform of the gravity data over a sphere. Lines are drawn at the extrema of the wavelet coefficients, which intersect at the anomaly source location. (b) Gravity anomaly due to a sphere with a radius of 2 km buried at a depth of 5 km having its center at (5, 5).

Acknowledgments

I am thankful to the editors of this volume, Prof. E. Chandrasekhar, Prof. V.P. Dimri, and Prof. V.M. Gadre for the invitation to contribute a chapter. I am greatly indebted to Prof. V.P. Dimri for his continuous support and encouragement and many fruitful discussions on the topic given in this chapter.

References

Berry, M.V., and Lewis, Z.V. 1980. On the Weistrass Mandelbrot fractal function. *Proceedings of the Royal Society of London* A370, 459–484.

Blackman, R.B., and Tukey, J.W. 1958. *The Measurement of Power Spectra*. Dover Publications, New York.

Brillouin, L. 1962. *Science and Information Theory*. 2nd ed., Academic Press, Inc. (Chapter 8), New York.

Caccia, D.C., Percival, D., Cannon, M.J., Raymond, G., and Bassingthwaighte, J.B. 1997. Analyzing exact fractal time series: Evaluating dispersional analysis and rescaled range methods. *Physica A* 246, 609–632.

Chamoli, A., Srivastava, R.P., and Dimri, V.P. 2006. Source depth characterization of potential field data of Bay of Bengal by continuous wavelet transform. *Indian Journal of Marine Sciences* 35(3), 195–204.

Dimri, V.P. 1992. *Deconvolution and Inverse Theory*. Elsevier Science Publishers, Amsterdam p. 230.

Dimri, V.P. 1998. Fractal behavior and detectibility limits of geophysical surveys. *Geophysics* 63, 1943–1946.

Dimri, V.P. (ed.) 2000. *Application of Fractals in Earth Sciences*. 238. A.A. Balkema, The Netherlands.

Dimri, V.P., and Vedanti, N. 2005. Scaling evidences of thermal properties in earth's crust and its implications. In *Fractal Behaviour of the Earth System*, edited by V.P. Dimri, 207. Springer, New York.

Dimri, V.P., Vedanti, N., and Chattopadhyay, S. 2005. Fractal analysis of aftershock sequence of Bhuj earthquake—a wavelet based approach. *Current Science* 88(10), 1617–1620.

Dimri, V.P., Srivastava, R.P., and Vedanti, N. 2011. Fractals and chaos. In *Encyclopedia of Solid Earth Geophysics*, edited by H. Gupta, Springer, The Netherlands.

Dimri, V.P., Srivastava, R.P., and Vedanti, N. 2012. *Fractal Models in Exploration Geophysics*. 165. Elsevier Science Publishers, Amsterdam.

Dmowska R., and Saltzman, B. (eds.) 1999. Advances in geophysics. *Long Range Persistence in Geophysical Time Series*. 40, 1–87. Academic Press, San Diego.

Hardy, H.H., and Beier, R.A. 1994. *Fractals in Reservoir Engineering*. World Scientific Publishing Company, Singapore.

Hewett, T.A. 1986. Fractal distribution of reservoir heterogeneity and their influence on fluid transport. SPE 15386, 61st Annual Technical Conference and Exhibition of the SPE, New Orleans, Louisiana.

Higuchi, T. 1990. Relationship between the fractal dimension and the power law index for a time series: A numerical investigation. *Physica D* 46, 254–264.

Hornby, P., Boschctti, F., and Horovitz, F.G. 1999. Analysis of potential field data in the wavelet domain. *Geophysical Journal International* 137, 175–196.

Hurst, H.E., Black, R.P., and Simaika, Y. M. 1965. *Long-term Storage*. Constable, London.

Keating, P. 1993. The fractal dimension of gravity data sets and its implications for gridding. *Geophysical Prospecting* 41, 983–993.

Korvin, G. 1992. *Fractal Models in the Earth Sciences*. Elsevier Science Publishers, Amsterdam.

Korvin, G., Boyd, D.M., and Odow, R. 1990. Fractal characterization of the South Australian gravity station network. *Geophysical Journal International* 100, 535–539.

Li, X.-G., Sacchi, M.D., and Ulrych, T.J. 1996. Wavelet transform inversion with a priory scale information. *Geophysics* 61(5), 1379–1385.

Lovejoy, S., Schertzer, D., and Ladoy, P. 1986. Fractal characterization of inhomogeneous geophysical measuring networks. *Nature* 319, 43–44.

Mandelbrot, B.B., and van Ness, J.W. 1968. Fractional Brownian motions, fractional noises and applications. *SIAM Review* 10, 422–437.

Mauri, G., Williams-Jones, G., and Saracco, G. 2011. MWTmat—application of multi-scale wavelet tomography on potential field. *Computers and Geosciences* 37, 1825–1835.

Maus, S., and Dimri, V.P. 1994. Fractal properties of potential fields caused by fractal sources. *Geophysical Research Letters* 21, 891–894.

Maus, S., and Dimri, V.P. 1995a. Basin depth estimation using scaling properties of potential fields. *Journal of Association of Exploration Geophysicists* 16, 131–139.

Maus, S., and Dimri, V.P. 1995b. Potential field power spectrum inversion for scaling geology. *Journal of Geophysical Research* 100, 12605–12616.

Maus, S., and Dimri, V.P. 1996. Depth estimation from the scaling power spectrum of potential field? *Geophysical Journal International* 124, 113–120.

Moreau, F., Gibert, D., Holschneider, M., and Saracco, G. 1997. Wavelet analysis of potential fields. *Inverse Problems* 13, 165–178.

Moreau, F., Gibert, D., Holschneider, M., and Saracco, G. 1999. Identification of sources of potential fields with the continuous wavelet transform: Basic theory. *Journal of Geophysical Research* 104(B3), 5003–5013.

Pilkington, M., and Todoeschuk, J.P. 1993. Fractal magnetization of continental crust. *Geophysical Research Letters* 20, 627–630.

Pilkington, M., Todoeschuk, J.P., and Gregotski, M.E. 1994. Using fractal crustal magnetization models in magnetic interpretation. *Geophysical Prospecting* 42, 677–692.

Reid, A.B. 1980. Aeromagnetic survey design. *Geophysics* 45, 973–976.

Sailhac, P., Galdeano, A., Gibert, D., Moreau, F., and Delor, C. 2000. Identification of sources of potential fields with the continuous wavelet transform: Complex wavelets and application to aeromagnetic profiles in French Guiana. *Journal of Geophysical Research* 105(B8), 19455–19475.

Sen, M.K., and Stoffa, P.L. 1995. *Global Optimization Methods in Geophysical Inversion*. Elsevier Science B.V. Amsterdam, The Netherlands.

Spector, A., and Grant, F.S. 1970. Statistical models for interpreting aero magnetic data. *Geophysics* 35, 293–302.

Srivastava, R.P., and Sen, M.K. 2009. Fractal based stochastic inversion of poststack seismic data using very fast simulated annealing. *Journal of Geophysics and Engineering* 6, 412–425.

Srivastava, R.P., and Sen, M.K. 2010. Stochastic inversion of prestack seismic data using fractal based prior. *Geophysics* 75(3), R47–R59.

Srivastava, R.P., Vedanti, N., and Dimri, V.P. 2007. Optimum design of a gravity survey network and its application to delineate the Jabera–Damoh structure in the Vindhyan basin, Central India. *Pure and Applied Geophysics* 164, 2009–2022.

Turcotte, D.L. 1997. *Fractals and Chaos in Geology and Geophysics*. 398. Cambridge University Press, Cambridge.

Vedanti, N., Srivastava, R.P., Sagode, J., and Dimri, V.P. 2005. An efficient nonlinear Occam's inversion algorithm with analytically computed first and second derivatives for DC resistivity sounding. *Computers and Geosciences* 31, 319–328.

6

Role of Multifractal Studies in Earthquake Prediction

S. S. Teotia and Dinesh Kumar

CONTENTS

Introduction 177
Similarity and Scaling... 178
Fractal Dimension 179
Multifractals and Multifractal Analysis... 180
Multifractal Analysis of Seismicity... 184
Conclusions... 190
References.. 191

Introduction

The concept of fractals was developed by Mandelbrot (1982) to bring together, under one heading, a large number of objects that contained structures nested within one another like Chinese boxes or Russian dolls. For example, the Sierpensky gasket consists of triangles and so on to the finest level. Mandelbrot (1967) introduced this concept into a geological context. Noting that the length of a rocky coastline increased as the length of a measuring rod decreased according to power law, and thus, the power law was associated with the fractal (fractional) dimension. Fractal geometry plays a vital role in describing the geometry of deterministic chaos and it can also describe the geometry of mountains, clouds, and galaxies. The basis of this observation is that rocky coastlines have a similar appearance when viewed at different scales (called scale-invariant distribution) and such features or phenomenon creating these scale-invariant features can be studied using the concept of fractals.

There are several geometries/objects in nature that cannot be described by Euclidean geometry and, therefore, it becomes difficult to fit them in Euclidean geometries/objects having integer dimensions such as 0, 1, 2, and 3. To understand the geometry of such objects, Mandelbrot (1967, 1982) came up with the concept of fractal geometry. Mandelbrot (1982) illustrated that the geometry of some irregular and fragmented objects present in nature are neither describable by Euclidean geometry nor by its two well-known generalizations, that is, Riemannian and Lobachevskian geometries. Examples of such geometries include the shape of a cloud, a mountain, a coastline, or a tree. Clouds are not spheres, mountains are not cones, coastlines are not circles, and bark is not smooth, nor does lightning travel in a straight line. The existence of these patterns was termed by Euclid as formless or "amorphous" and this challenged Mandelbrot to look into the unexplored domain of the morphology of the amorphous. Nature, therefore, exhibits not simply a higher degree but an altogether different level of complexity.

Mandelbrot (1982) subsequently developed a new geometry of nature, and implemented its use in a number of diverse fields, although its root could be traced to Cantor, Koch, Peano, and Hausdorff. It describes many of the irregular and fragmented patterns around us, and leads to fully-fledged theories by identifying families of shapes called fractals. The most useful fractal in nature involves chance and both their regularity and irregularities are statistical. Also, the shapes described by fractals tend to be scaling, implying that the degree of their irregularity (fragmentation) is self-similar at all scales. The concept of fractal (Hausdorff) dimension plays a central role in the theory of fractals.

A fractal is defined as a set for which the Hausdorff–Bescovitch dimension (D) strictly exceeds the topological dimension (D_T). Every set with a noninteger D is a fractal. For example, the original Cantor set, which is a fractal because $D = \log 2/\log 3 = 0.6309 > 0$, whereas $D_T = 0$. The Cantor set can be tailored and generalized so that $D_T = 0$, whereas D takes on any desired value. The original Koch curve is a fractal because $D = \log 4/\log 3 = 1.2618 > 1$, whereas $D_T = 1$. However, a fractal may also have an integer D. As an example, the trail of Brownian motion is a fractal because $D = 2$, whereas $D_T = 1$.

Several of the above listed values of D are fractional, and indeed the Hausdorff–Bescovitch dimension is often called fractional dimension. Mandelbrot termed D as fractal dimension. The topological dimensions (D_T) of all single-island coastlines have the same form because they are topologically identical to a circle. The topological dimension is the same for all coastlines and circles (equal to 1). The topology fails to discriminate between different coastlines. However, different coastlines tend to have different fractal dimensions. Differences in fractal dimensions express differences in a nontopological aspect of form, which Mandelbrot (1975) proposed to call *fractal form*.

Similarity and Scaling

In the foregoing, we reviewed aspects of fractals, which have some interesting properties. For example, if we enlarge a part of a fractal image, it looks very much like the whole image without any loss of resolution or the details. This underlying property of fractals is called self-similarity. Self-similarity is related to and extends one of the most fruitful notions of elementary geometry, that is, the notion of similarity. Two objects are similar if they have the same shape, regardless of their size. Corresponding angles, however, must be equal, and corresponding line segments must all have the same factor of proportionality. For example, when a photograph is enlarged, it is enlarged by the same factor in both horizontal and vertical directions. We call this enlargement factor scaling. The above transformation between objects is called similarity transformation because the property of objects whereby magnified subsets appear similar or identical to the whole and to each other is known as similarity. It sets fractals apart from Euclidean shapes, which generally become smoother under such transformations. Thus, fractal shapes are self-similar and scale-invariant, that is, they possess no characteristic size. The property of self-similarity, or scaling, is one of the central concepts of fractal geometry.

Fractal Dimension

For the estimation of fractal dimension, we try to establish power law between two quantities, that is, measured quantity (i.e., Y) on the one hand, and the scale (X) on the other hand, which do not vary arbitrarily but are related by rule, a rule that allows us to compute one quantity from the other. This, as has been explained earlier, is a power law of the form $Y \propto X^d$. This law has become the basis for the analysis of fractal dimension. Dimension is a subtle concept. At the turn of the century, one of the major problems in mathematics was to determine what dimension meant and which properties it had. Since then, the situation has become somewhat worse because mathematicians have come up with several different notions of dimensions: topological dimension, Hausdorff dimension, fractal dimension, self-similarity dimension, capacity dimension, information dimension, correlation dimension, etc., are all related. However, each of them has greater applicability in specific areas of applications. Sometimes, they all make sense and are the same, and in some cases, they all make sense but do not agree (Saupe et al. 1992).

Of these notions of dimensions, the self-similarity, capacity, information, and correlation dimensions have the most applications in science. Some of these dimensions are defined as follows (Grassberger and Procraccia 1983; Takayasu 1990):

1. Similarity dimension D_s is defined for an exactly self-similar set as $D_s = \log \beta / \log \alpha$, where α is linear size and β is the number of similar objects.
2. Capacity dimension D_c defined as $D_c = \log N(\varepsilon) / \log (1/\varepsilon)$, where N is the smallest number of coverings of the set with size ε.

3. Information dimension D_1 is defined for probabilistic distribution as

$$D_1 = \lim_{\varepsilon \to 0} \frac{\sum_{i=0}^{N(s)} P_i(\varepsilon) \log P_i(\varepsilon)}{\log \varepsilon}$$

where $P_i(\varepsilon)$ is the probability for a point to belong to the ith box of size ε.

4. Correlation dimension D is defined from the correlation integral $C(r)$ by the relation,

$$C(r) = \frac{1}{N^2} \sum_{\substack{i,j=1 \\ i \neq j}}^{N} H\left(r - \left|X_i - X_j\right|\right)$$

where H is the Heaviside step function

$$H(x) = 0 \text{ if } x < 0$$

$$= 1 \text{ if } x > 0$$

$C(r)$ is the correlation integral, r is the scaling radius, and N is the total number of data points within the search region in a certain time interval (also termed as the sample volume) and X_i, X_j are points in space.

The relation between the correlation dimension and the cumulative correlation function is based on the power law.

$$C(r) = r^{D_{cor}}$$

$$\log C(r) = D_{cor} \log r$$

The determination of the correlation dimension is found by plotting $C(r)$ versus r on a log-log graph. The region in which the power law is obeyed appears as a straight line that is called the scaling region. The slope (which is an estimate of correlation dimension) is formed by fitting a least squares line in the scaling region. The correlation dimension depends on the number of data points. The number of data points used in analysis is called sample volume.

Multifractals and Multifractal Analysis

Fractals exhibit a fractal structure over a wide range of fracture scales, that is, from the scales of microfractures to megafaults (Aki 1981; Allegre et al. 1982;

Brown and Scholz 1985; Scholz 1986; Scholz and Aviles 1986; Turcotte 1986a,b; Okubo and Aki 1987; Aviles et al. 1987; Hirata et al. 1987; Hirata 1987). Earthquakes belong to a class of spatiotemporal point processes, marked by magnitude. Consider a seismotectonic map of a region and the measure η on such a map could be seismicity distribution of earthquakes from distribution of fractures/faults. In each subset S of the map, the measure attributes a quantity $\eta(S)$, which is the density distribution of fractures/faults below S in the subsurface up to some prescribed level. On dividing the map into two equally sized subregions, S_1 and S_2, we may find that $\eta(S_1)$ will not be the same as $\eta(S_2)$ as is the case with seismicity of two different tectonic regions. If S_1 is further subdivided into two equally sized pieces S_{11} and S_{22}, their seismicity distribution may again differ in the further subdivided regions. The subdivision could be carried through the joints in rocks, in which some joints develop into fractures/faults. The densities of fractures/faults give rise to epicenters of earthquakes in some tectonic regions. Therefore, the distribution of earthquake epicenters (i.e., seismicity) could be considered as a measure of η. There could be number of other quantities that may exhibit the same behavior, such as fluid turbulence and amount of ground water below S (Saupe et al. 1992), which means that the quantity η is an example of a measure that is irregular at all scales. When irregularity is the same at all scales, or at least statistically the same, one says that the measure is self-similar, that is, multifractal. Multifractal could be seen as an extension of fractals. The topic of multifractals is bound to become of increasing importance to geophysics. Mandelbrot (1989) has described the closely related notions of self-similar fractals or self-similar multifractals, which can be phrased in many ways, but the geophysicist might be best able understand in the context of the distribution of a rare mineral, such as copper. A multifractal object is more complex in the sense that it is always invariant by translation.

The spatial distributions of epicenters or hypocenters of earthquakes are known to be fractal, and fractal dimensions in several regions have been estimated (Kagan and Knopoff 1980; Sadvskiy et al. 1984; Hirata 1989a,b). In such systems, the number of fractures that are larger than a specified size are related by power law in size. The physical laws governing the fractal structures are scale-invariant in nature because the occurrence of earthquakes is causally related to the fractures, which have a fractal structure in their space, time, and magnitude distributions. Fractal dimension is a good parameter to characterize the spatial distribution of point sets such as earthquake hypocenters, especially the degree of clustering. The generalization from fractal sets to multifractal measures involves the passage from objects that are characterized primarily by one number, namely, a fractal dimension, to objects that are characterized by a function (Mandelbrot 1989). Most fractals in nature and in dynamic systems are known to be heterogeneous. To define heterogeneous fractals, a unique fractal dimension is not sufficient to characterize them. Such fractals are called multifractals and they are characterized by the generalized dimension D_q or $f(\alpha)$ spectrum (Hentschel and Procraccia 1983; Halsey et al. 1986; Hirabayashi et al. 1992).

$$D_q = \left[\frac{1}{q-1}\lim_{r\to 0}\frac{\log\left(\sum_i\{P_i(r)\}^q\right)}{\log r}\right] \tag{6.1}$$

where P_i is the probability that events fall into the box of length r during the process of fragmentation. The generalized dimension is more useful for the comprehensive study of turbulent phenomenon in nature and dynamic systems, which are heterogeneous fractals. The multifractal dimension, that is, generalized dimension D_q is a parameter representing the complicated fractal structure or multiscaling nature of nonlinear processes. The general methods for calculating D_q are fixed-mass method, the fixed-radius method, and the box counting method (Greenside et al. 1982; Grassberger and Procraccia 1983; Halsey et al. 1986). The box counting method, fixed-mass, and fixed-radius methods work well; provided the number data are very large. The relationship between the generalized dimension D_q and singularity spectrum $f(\alpha)$ is expressed by (Halsey et al. 1986; Hirabayashi et al. 1992)

$$D_q = \frac{1}{q-1}\left[q\alpha(q) - f(\alpha(q))\right] \tag{6.2}$$

This is deduced from Equation 6.1 using the method of steepest descent. The revised expression is the Legendre transformation of Equation 6.2, such that

$$\alpha(q) = \frac{d}{dq}\left[(q-1)D_q\right],$$

$$f(\alpha(q)) = q\alpha(q) - (q-1)D_q \tag{6.3}$$

The fractal structures may have either homogeneous or multiscaling fractal sets. Fractal sets with multiscaling are heterogeneous fractal sets and are called multifractal sets. Most fractals in nature are said to be heterogeneous (Stanley and Meakin 1988; Mandelbrot 1989). Such fractals are characterized by a generalized dimension D_q, where q takes the value ranging from $-\infty\ldots -2, -1, 0, 1.1, 2,\ldots \infty$. A variety of natural phenomena are multifractal, the multifractal property of such phenomenon in space and time must be compared using the whole spectrum of D_q or $f(\alpha)$. The graph of D_q versus q

is called the D_q spectra. It has been found that homogeneous fractals (mono-fractal) have a flat D_q spectrum whereas heterogeneous fractals lead to a variation in D_q spectra.

Recent studies have shown that many natural phenomena such as the spatial distribution of earthquakes (Telesca et al. 2003a,b, 2004, 2005) and fluid turbulence are heterogeneous fractals (Mandelbrot 1989). For heterogeneous fractals, a unique fractal dimension is not sufficient to characterize them and it differs by the method used to calculate it. Such fractals are called multifractals and they are characterized by generalized D_q or the $f(\alpha)$ spectrum (Hentschel and Procraccia 1983; Halsey et al. 1986). A wide variety of heterogeneous phenomena such as diffusion-limited aggregation, dendritic crystallization, dielectric breakdown, various fingering, and river flow multifractals and multifractal analysis must be used to characterize these complex phenomena (Stanley and Meakin 1988). If the spatiotemporal distribution of earthquakes is multifractal, the fractal property of earthquakes in different areas and different time spans must be compared using whole spectra of D_q or $f(\alpha)$ (Hirabayashi et al. 1992). Several studies have been performed to investigate the temporal variation of heterogeneity in seismicity using multifractal analysis in various seismic regions (Hirata and Imoto 1991; Hirabayashi et al. 1992; Li et al. 1994; Dimitriu et al. 2000; Teotia et al. 1997; Teotia and Kumar 2007, 2011; Telesca et al. 2005). Recently, it has been recognized that the parameters which describe seismicity in a region show a spatial and temporal evolution that may be associated with the process of generating large-sized/great-sized earthquakes, that is, the evolution of fractures from microfractures to megafaults (Telesca et al. 2003b,c). The change in seismicity pattern of earthquakes is reflected in the generalized dimension D_q and D_q spectra of the seismicity. Therefore, the study of temporal variation D_q and D_q spectra may be used to study the changes in the seismicity structure before the occurrence of large earthquakes, which may prove to be useful in understanding the preparation zone of large-sized to great-sized earthquakes. Accordingly, Hirata and Imoto (1991) have performed multifractal analysis of microearthquake data from the Kanto region using a correlation integral method. Hirabayashi et al. (1992) have performed multifractal analyses of seismicity in regions of Japan, California, and Greece using various procedures such as fixed mass and fixed radius methods. Li et al. (1994) have performed multifractal analyses of the spatial distribution of earthquakes from Tanshan region ($Ml > 1.8$) using an extended Grassberger–Procraccia method of dimension D_q estimation. Teotia et al. (1997), Dimri (2000), Teotia and Kumar (2007), and Sunmonu et al. (2001) have studied the multifractal characteristics of seismicity of Himalaya region and Sumatra region ($m_b \geq 4.5$). Telesca et al. (2003a) have performed monofractal and multifractal characterizations of geoelectrical signals measured in southern Italy. They observed that multifractal parameters revealed good discrimination between the geoelectrical signals measured in seismic areas and those recorded in aseismic areas. Telesca et al. (2005) have also studied multifractal variability in geoelectrical signals and correlations with seismicity in a nice case study from southern Italy. In that study, they analyzed multifractal fluctuations in the time dynamics of geoelectrical data using the multifractal–detrended fluctuation analysis (MF-DFA), which detects multifractality as well as the time evolution of the multifractality, suggesting that the multifractal degree increases before the occurrence of earthquakes. This study aims to propose another approach to investigate the complex dynamics of earthquake-related geoelectrical signals. Enescu et al. (2004) have presented a wavelet-based multifractal analysis of real and simulated time series of earthquakes. Their study introduces a new approach that is based on the continuous wavelet transform modulus maxima method for qualitatively and quantitatively describing the complex patterns of seismicity; in particular, their multifractal and clustering properties. The result of their study revealed several distinct scaling domains, which indicate different periodic trends of the time series. Roy and Padhi (2007) have studied the multifractal characteristics of earthquakes from the Bam region in southeastern Iran, and have found the relationship between spatiotemporal pattern of seismicity and multifractal characteristics before the occurrence of a large-sized Iran earthquake. The October 8, 2005, Muzaffrabad–Kashmir earthquake ($Mw = 7.6$) killed more than 80,000 people and was the deadliest in the history of the Indian subcontinent. The earthquake occurred on a rupture plane 75 km long and 35 km wide, with a strike of 331° and a dip angle of 29°. The epicenter of the event was located north of Muzaffrabad (34.493°N and 73.629°E) in Kashmir Himalaya (USGS). Teotia and Kumar (2011) deal with the multifractal analysis of the seismicity in the region, which resulted in the Muzaffarabad–Kashmir earthquake.

Multifractal Analysis of Seismicity

For the analysis of seismicity in northwestern Himalaya, the earthquake data set of USGS for the grids (30°N, 72°E), (30°N, 77°E), (36°N, 77°E), (36°N, 72°E) and (28°N, 77°E), (28°N, 85°E), (35°N, 85°E), (35°N, 77°E) were analyzed. The tectonic map, along with grids, is shown in Figure 6.1. The time window for analysis was chosen to be 1973 to 2009. To gain an unbiased and homogeneous data set, we restricted the data by setting a lower limit and the analysis time period and it was found to be complete for the time period 1973 to 2009 for magnitude threshold $m_b \geq 4.5$. The USGS data set used in this study is complete for $m_b \geq 4.5$. The study area data set covers two merged regions, that is, (30°N, 72°E), (30°N, 77°E), (36°N, 77°E), (36°N, 72°E) and (28°N, 77°E), (28°N, 85°E), (35°N, 85°E), (35°N, 77°E). The minimum and maximum latitude and longitude of the study area are (28°N, 36°N) and (72°E, 85°E). The seismicity map and number of earthquakes occurring per year for $m_b \geq 4.5$

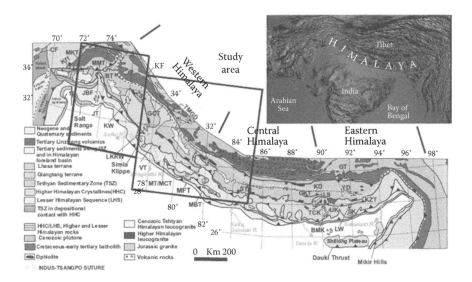

FIGURE 6.1
The tectonic map and study area. (Tectonic map modified from Yin, A., *Earth Sci. Rev.* 76, 1–131, 2006. Yin, A. et al., *Geol. Soc. Am. Bull.* 122, 360–395, 2010. Study area from Teotia, S.S., and Kumar, D., *Nonlinear Process. Geophys.* 18, 111–118, 2011.)

in the selected region for the time period 1973 to 2009 are shown in Figure 6.2a and b.

The available methods for calculating D_q are the box counting, the fixed-radius, and the fixed-mass methods (Greenside et al. 1982; Grassberger et al. 1988; Baddi and Broggi 1988). These methods work well, provided the number of data points are very large. The extended Grassberger and Procaccia method was used to calculate the generalized dimension D_q and D_q spectrum (Grassberger and Procaccia 1983). This method has been applied by many authors on earthquake data from different regions for the estimation of generalized dimension D_q even for small data points, that is, 100. It is described as

$$\log C_q(r) = D_q \log(r)(r \to 0) \tag{6.4}$$

$$C_q(r) = \lim_{r \to 0} \left[\frac{1}{N} \sum_{j=1}^{N} \left\{ \frac{1}{N} \sum_{\substack{i=1 \\ i \neq j}}^{N} H\left(r - |X_i - X_j|\right) \right\}^{q-1} \right]^{1/q-1} \tag{6.5}$$

where r is the scaling radius, N is the total number of data points within a search region in a certain time interval (also called the sample volume): X_i is the epicentral location (given in latitude and longitude) of the ith event, X_j is

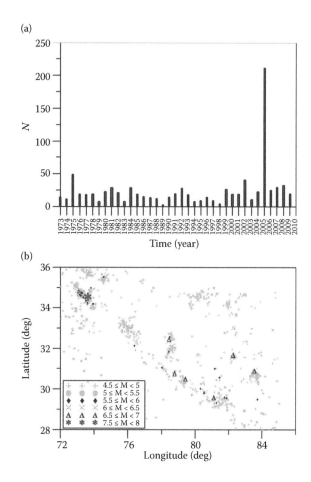

FIGURE 6.2
(a) Year-wise distribution of the number of earthquakes from 1973 to 2009. (b) Seismicity of the selected region.

the epicenter (given in latitude and longitude) of the jth event, $C_q(r)$ is the qth integral, and $H(.)$ is the Heaviside step function.

For estimating D_q spectra as a function of time, a time series of earthquake epicenters has to be formed and divided into a subseries (subsets). Let set $\{X_i, M_i\}$, $i = 1, M$ be a complete set of earthquakes occurring in time period analyzed, and M_i the magnitude of an earthquake occurring at time t_i. Thus, the earthquake constitutes a time series of N elements. The time series consists of 902 events in the region. We consider this time series as the original data set for multifractal analysis of this region. In this study, the original data set for northwestern Himalaya is divided into 27 subsets (i.e., $S_1 - S_{27}$). Each subset in the region consists of 100 events with an overlap 70 events. The shift of 30 events has been used in the analysis. The number of events in each subset ($N = 100$) is large enough to provide reliable estimates of b value using the maximum likelihood method (Utsu 1965). The correlation integral is calculated using Equation 6.5 for the epicentral distribution X_i of the subset. The distance r between subsets is calculated by using a spherical triangle (Bullen and Bolt 1985; Hirata 1989a,b). For epicentral distribution with a fractal structure, the following power law relationship is obtained in the scaling region:

$$C_q(r) \sim r^{D_q}$$

An appropriate scaling region has been estimated before the computation of generalized dimension D_q. Figure 6.3 shows the scaling region, which is a linear segment in the graph of log $C_q(r)$ versus log r, that has been presumed linear in the range of log r(−0.2 to 0.7), that is, the first seven points in log $C_q(r)$ versus the log r plot. The generalized dimensions for all 27 subsets have been estimated by least square fit from the first seven points (Table 6.1).

The observation of seismicity suggests a relationship between the distribution of earthquake magnitudes and the distribution of earthquake epicenters (Hirata 1989a,b; Henderson et al. 1984). The fractal dimension D_2 and D_{-2} for all 27 subsets are shown in Figure 6.4 and Table 6.1. Figure 6.4 shows the spatiotemporal variation of generalized fractal dimension D_q ($q = 2, -2$) along with b values in all 27 subsets analyzed for the time period 1973 to 2009 in the selected region. We note that the changes in b values as well as in D_2 starts from subset S_{15} (1995.544–2001.759), with the lowest value in subset S_{21} (2004.773–2004.773). The Muzaffrabad–Kashmir large-sized earthquake lies in subset 21. It is evident that there was a significant increase in clustering

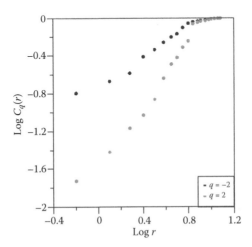

FIGURE 6.3
log $C_q(r)$ versus log r for $q = 2, -2$ for subset S_1.

TABLE 6.1

Subsets' Descriptions Along with Time Periods and Fractal Dimensions

Serial No.	Subset Name	Subset Event	Time period (years)	D_{-2}	D_2	$M \geq 7.0$ (No. of Events)
1	S_1	1–100	1972.044–1976.137	1.4790 ± 0.0207 (SD)	0.7429 ± 0.0067 (SD)	–
2	S_2	31–130	1974.052–1977.745	1.5459 ± 0.0269 (SD)	0.8087 ± 0.0080 (SD)	–
3	S_3	61–160	1974.556–1979.645	1.7412 ± 0.0385 (SD)	0.9738 ± 0.0094 (SD)	–
4	S_4	91–190	1975.688–1980.863	1.7830 ± 0.0876 (SD)	1.0728 ± 0.0110 (SD)	–
5	S_5	121–220	1977.323–1982.784	1.8278 ± 0.1242 (SD)	1.0081 ± 0.0095 (SD)	–
6	S_6	151–250	1979.525–1983.992	1.7651 ± 0.0389 (SD)	0.9824 ± 0.0089 (SD)	–
7	S_7	181–280	1980.701–1985.573	1.7062 ± 0.0187 (SD)	1.0439 ± 0.0168 (SD)	–
8	S_8	211–310	1981.956–1987.921	1.5712 ± 0.0038 (SD)	1.1610 ± 0.0054 (SD)	–
9	S_9	241–340	1983.708–1990.403	1.5898 ± 0.1266 (SD)	1.1835 ± 0.0039 (SD)	–
10	S_{10}	271–370	1985.162–1991.648	1.7108 ± 0.0516 (SD)	1.0779 ± 0.0059 (SD)	–
11	S_{11}	301–400	1987.063–1993.586	1.6417 ± 0.0511 (SD)	1.1652 ± 0.0045 (SD)	–
12	S_{12}	331–430	1990.036–1996.014	1.6507 ± 0.0555 (SD)	1.1636 ± 0.0046 (SD)	–
13	S_{13}	361–460	1991.309–1998.326	1.6790 ± 0.0219 (SD)	1.0593 ± 0.0053 (SD)	–
14	S_{14}	391–490	1992.545–2000.096	1.7828 ± 0.0712 (SD)	1.1022 ± 0.0029 (SD)	–
15	S_{15}	421–520	1995.544–2001.759	1.7543 ± 0.0373 (SD)	1.0950 ± 0.0028 (SD)	–
16	S_{16}	451–550	1998.247–2002.170	1.6841 ± 0.0221 (SD)	0.8221 ± 0.0095 (SD)	–
17	S_{17}	481–580	1999.544–2003.833	1.6731 ± 0.0484 (SD)	0.6972 ± 0.0031 (SD)	–
18	S_{18}	511–610	2001.145–2004.773	1.6513 ± 0.0716 (SD)	0.5766 ± 0.0018 (SD)	–
19	S_{19}	541–640	2001.863–2004.773	1.5511 ± 0.1167 (SD)	0.4192 ± 0.0018 (SD)	–
20	S_{20}	571–670	2003.527–2004.773	1.2007 ± 0.0298 (SD)	0.1505 ± 0.0002 (SD)	–
21	S_{21}	601–700	2004.773–2004.773	0.0142 ± 0.0000 (SD)	0.0111 ± 0.0000 (SD)	–
22	S_{22}	631–730	2004.773–2004.775	0.1539 ± 0.0056 (SD)	0.0371 ± 0.0003 (SD)	1
23	S_{23}	661–760	2004.773–2004.803	0.1629 ± 0.0063 (SD)	0.0635 ± 0.0009 (SD)	1
24	S_{24}	691–790	2004.773–2004.970	0.8422 ± 0.0707 (SD)	0.1121 ± 0.0020 (SD)	1
25	S_{25}	721–820	2004.775–2006.049	1.3293 ± 0.0491 (SD)	0.1622 ± 0.0010 (SD)	–
26	S_{26}	751–850	2004.792–2007.188	1.6297 ± 0.1409 (SD)	0.3845 ± 0.0009 (SD)	–
27	S_{27}	781–880	2004.890–2007.902	1.6696 ± 0.1675 (SD)	0.6576 ± 0.0081 (SD)	–

Source: Teotia, S.S., and Kumar, D., *Nonlinear Process. Geophys.* 18, 111–118, 2011.

before the occurrence of the Muzaffrabad–Kashmir earthquake. The same is evident in $D_q - q$ spectra. The flat $D_q - q$ spectra in subset 21 shows the convergence of epicenters to a point, that is, confinement of seismicity in the zone of preparation of the large-sized Muzaffrabad–Kashmir earthquake ($Mw = 7.6$) leading to very small values of generalized dimension. Subset 21 includes the epicenters of the aftershock events of the Muzaffrabad–Kashmir earthquake of October 8, 2005.

The steep slope in D_q spectra also provides a better indication of the preparation zone for large-sized earthquakes in the Himalayan region, which is considered to be a storehouse of elastic energy and has the potential to

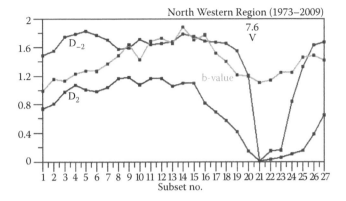

FIGURE 6.4
The temporal variation of generalized dimension $D_q(q = 2, -2)$ along with the temporal variation of b value. (After Teotia, S.S., and Kumar, D., *Nonlinear Process. Geophys.* 18, 111–118, 2011.)

generate moderate, large-sized, and great-sized earthquakes. The same has been observed for the northwestern Himalaya region, which showed a steep slope in D_q spectra from subset S_{15} (1995.544–2001.759) to subset S_{20} (2003.527–2004.773) before the occurrence of the large-sized ($Mw = 7.6$) Muzaffrabad–Kashmir earthquake (Figure 6.5). The steep slope in $D_q - q$ spectra relates to clustering in the zone of preparation or nucleation. Consistent and significant decreases in D_q and b values were observed before the occurrence of the Muzaffrabad–Kashmir earthquake. The consistent decrease in spatial dimension (D_s) might correspond to the introduction of clustering of seismicity in the region, showing the region's preparedness for the occurrence of major-sized events. De Rubeis et al. (1993) also found the temporal evolution of three seismic zones from Italy using a correlation integral fractal method and the decrease in fractal dimension before the occurrence of major events. Their study for Italian regions and our study for Himalayan regions show the similar behavior in spatiotemporal variation of fractal dimension as the seismicity in the region evolves from extended distribution of epicenters to clustered distribution, and again back to distributed seismicity. This is evident from the decrease in D_q from subset 15 to subset 21, and an increase in D_q from subset 21 to subset 27 (Figure 6.4). Therefore, the spatiotemporal evolution of seismicity can be best understood in the framework of self-organized criticality (SOC). The concept of SOC, which was introduced by Bak and Tang (1989), can be applied to the distribution of seismicity. SOC can be defined as a natural system that is in a marginally stable state, and when perturbed from this state, will evolve back to the state of marginal stability. A simple cellular–automata model illustrates SOC. The results of this study support the evolution of epicenter distribution from a random spatial distribution to an organized fractal structure such as that given in a modified SOC model (Ito and Matsuzaki 1990). Thus, the application of nonlinear

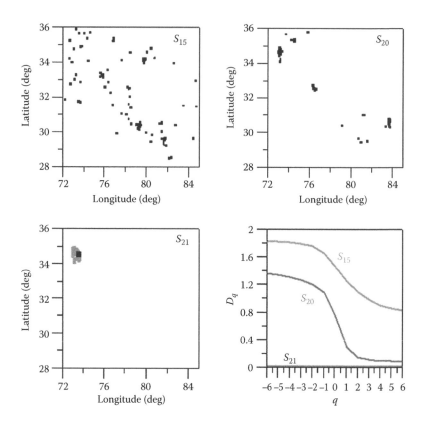

FIGURE 6.5
The seismicity of subsets S_{15}, S_{20} and S_{21}. The Muzaffrabad–Kashmir earthquake of October 2005 lies in subset S_{21}. The corresponding $D_q - q$ spectra of these subsets are also shown. (After Teotia, S.S., and Kumar, D., *Nonlinear Process. Geophys.* 18, 111–118, 2011.)

geophysics methods is important for extreme phenomena and new hazard assessment techniques (Rundle et al. 2003).

Conclusions

In general, the change in seismicity pattern of earthquakes is reflected in generalized dimension D_q and D_q spectra. The introduction of clustering leads to the heterogeneity of fractals, which leads to the steep slope in D_q spectra. The steep slope in D_q spectra is found to be indicative of significant clustering in the region. Temporal variations of heterogeneity in the seismicity using multifractal analysis have been reported in other tectonic regions for both microseismicity as well as macroseismicity data (Hirata and Imoto

1991; Hirabayashi et al. 1992; Li et al. 1994; Dimitriu et al. 2000; Telesca et al. 2005; Teotia and Kumar 2007; Teotia and Kumar 2011). Therefore, the study of temporal variation of D_q spectra may be used to study the changes in seismicity structure before the occurrence of large-size earthquakes may prove to be useful in understanding the preparation zone of large-sized to great-sized earthquakes.

References

Aki, K. A probabilistic synthesis of precursory phenomena. In *Earthquake Prediction: An International Review, Maurice Ewing Series* 4, edited by Simpson, D.W., and Richard, P.G. American Geophysical Union, Washington DC, 566–574, 1981.

Allegre, C.J., Le Mouel, J.L., and Provost, A. Scaling rules in rock failure and possible implications for earthquake prediction. *Nature* 297, 47–49, 1982.

Aviles, C.A., Scholz, C.H., and Boatwright, J. Fractal analysis applied to characteristics segments of the San Andreas Fault. *Journal of Geophysical Research* 92, 331–344, 1987.

Baddi, R., and Broggi, G. Measurement of the dimension spectrum f(α): Fixed-mass approach. *Physics Letters A* 131, 339–343, 1988.

Bak, P., and Tang, C. Self-organized criticality. *Physical Review A* 38, 364–371, 1989.

Brown, S.R., and Scholz, C.H. Broadband width with study of the topography of the natural surfaces. *Journal of Geophysical Research* 90, 12575–12582, 1985.

Bullen, K.E., and Bolt, B.A. *An Introduction to Theory of Seismology.* Cambridge University Press, New York, 1985.

De Rubeis, V., Dimitriu, P., Papadimitriou, E., and Tosi, P. Recurrence patterns in the spatial behaviour of Italian seismicity revealed by the fractal approach. *Geophysical Research Letters* 20(18), 1911–1914, 1993.

Dimitriu, P.P., Scordilis, E.M., and Karacostas, V.G. Multifractal analysis of the Arnea, Greece seismicity with potential implications for earthquake prediction. *Natural Hazards* 21, 277–295, 2000.

Dimri, V.P., ed. *Application of Fractals in Earth Sciences.* Balkema, A.A., 238, 2000.

Enescu B., Ito, K., and Struzik, Z.R. Wavelet-based multifractal analysis of real and simulated time series. *Annals, Disaster Prevention Research Institute, Kyoto University* 47B, 2004.

Grassberger, P., and Procraccia, I. Measuring the strangeness of strange attractors. *Physica D* 9, 189–208, 1983.

Grassberger, P., Baddi, R., and Politi, A. Scaling laws for invariant measures on hyperbolic attractors. *Journal of Statistical Physics* 51, 135–178, 1988.

Greenside, H.S., Wolf, A., Swift, J., and Pignataro, T. Impracticality of box-counting algorithm for calculating the dimensionality of strange attractors. *Physical Review A* 25, 3453–3459, 1982.

Halsey, T.C., Jensen, M.H., Kadanoff, L.P., Procraccia, I., and Shraiman, B.I. Fractal measures and their singularities: The characterizations of strange sets. *Physical Review A* 33, 1141–1151, 1986.

Hentschel, H.G.E., and Procraccia, I. The infinite number of generalized dimensions of fractals and strange attractor. *Physica D* 8, 435–444, 1983.

Henderson, J., Main, I.G., Pearce, R.G., and Takeya, M. Seismicity in north-eastern Brazil: Fractal clustering and the evaluation of *b* value. *Geophysical Journal International* 116, 217–226, 1994.

Hirata, T. Omori's power law aftershock sequences of microfracturing in rock fracture experiment. *Journal of Geophysical Research* 92, 6215–6221, 1987.

Hirata, T. Fractal dimension of fault systems in Japan: Fractal structure in rock fracture geometry at various scales. *Pure and Applied Geophysics* 131, 157–170, 1989a.

Hirata, T. A correlation between the *b* value and the fractal dimension of earthquakes. *Journal of Geophysical Research* 94, 7507–7514, 1989b.

Hirata, T., and Imoto, M. Multifractal analysis of spatial distribution of micro earthquakes in the Kanto region. *Geophysical Journal International* 107, 155–162, 1991.

Hirata, T., Satoh, T., and Ito, K. Fractal structure of spatial distribution of microfracturing in rock. *Geophysical Journal of the Royal Astronomical Society* 90, 369–374, 1987.

Hirabayashi, T., Ito, K., and Yoshii, T. Multifractal analysis of earthquakes. *Pure and Applied Geophysics* 138, 561–610, 1992.

Ito, K., and Matsuzaki, M. Earthquakes as self-organized critical phenomena. *Journal of Geophysical Research* 95(B5), 6853–6860, 1990.

Kagan, Y.Y., and Knopoff, L. Spatial distribution of earthquakes: The two-point correlation function. *Geophysical Journal of the Royal Astronomical Society* 62, 697–717, 1980.

Li, D., Zhaobi, Z., and Binghong, W. Research into the multifractal of earthquake spatial distribution. *Tectonophysics* 223, 91–97, 1994.

Mandelbrot, B.B. How long is the coast of Britain? Statistical self-similarity and fractional dimension. *Science* 156, 636–638, 1967.

Mandelbrot, B.B. Stochastic models for the earth's relief, the shape and the fractal dimension of the coastlines, and the number-area rule for islands. *Proceedings of the National Academy of Sciences of the United States of America* 72, 3825–3828, 1975.

Mandelbrot, B.B. *The Fractal Geometry of Nature.* Freeman, San Francisco, 468 p, 1982.

Mandelbrot, B.B. Multifractal measures, especially for geophysicists. *Pure and Applied Geophysics* 131, 5–42, 1989.

Okubo, P.G., and Aki, K. *Fractal Geometry of Nature.* W.H. Freeman, New York, 468 p, 1987.

Rundle, J.B., Turcotte, D.L., Shcherbakov, R., Klein, W., and Sammis, C. Statistical physics approach to understanding the multiscale dynamics of earthquake fault system. *Reviews of Geophysics* 41(4), 1019, 2003.

Roy, P.N.S., and Padhi, A. Multifractal analysis of earthquakes in the southeastern Iran Bam region. *Pure and Applied Geophysics* 164, 2271–2290, 2007.

Sadvskiy, M.A., Golubeva, T.V., Pisarenko, V.F., and Shnirman, M.G. Characteristic dimensions of rock and hierarchical properties of seismicity. *Izvestiya, Academy of Sciences, USSR: Physics of the Solid Earth* 20, 87–96, 1984.

Saupe, D., Jurgens, H., and Peitgen, O.-H., eds. *Chaos and Fractals.* Springer-Verlag, New York, 1992.

Scholz, C.H. Microfractures, aftershocks and seismicity. *Bulletin of the Seismological Society of America* 58, 1117–1130, 1986.

Scholz, C.H., and Aviles, C.A. The fractal geometry of faults and faulting. In *Earthquake Source Mechanics*, edited by Das, S., Boatwright, J., and Scholz, C., American Geophysical Union 37, 147–155, 1986.

Stanley, H.E., and Meakin, P. Multifractal phenomena in physics and chemistry. *Nature* 335, 405–409, 1988.

Sunmonu, L.A., Dimri, V.P., Ravi Prakash, M., and Bansal, A.R. Multifractal approach of the time series of M > 7 earthquakes in Himalayan region and its vicinity during 1985–1995. *Journal of the Geological Society of India* 58, 163–169, 2001.

Takayasu, H. *Fractals in Physical Sciences*. Manchester University Press, New York. 169, 1990.

Telesca, L., Colangelo, G., Lapenna, V., and Macchiato, M. Monofractal and multifractal characterization of geoelectrical signals measured in southern Italy. *Chaos, Solitons & Fractals* 18, 385–399, 2003a.

Telesca, L., Lapenna, V., and Macchiato, M. Investigating the time-clustering properties in seismicity of Umberia–Marche region (Central Italy). *Chaos, Solitons & Fractals* 18(2), 202–217, 2003b.

Telesca, L., Lapenna, V., and Macchiato, M. Spatial variability of the time-correlated behaviour in Italian seismicity. *Earth and Planetary Science Letters* 212(3–4), 279–290, 2003c.

Telesca, L., Lapenna, V., and Macchiato, M. Detrended fluctuation analysis of the spatial variability of the temporal distribution of southern California seismicity. *Chaos, Solitons & Fractals* 21(2), 335–342, 2004.

Telesca, L., Colangelo, G., and Lapenna, V. Multifractal variability in geoelectrical signals and correlations with seismicity: A study case in southern Italy. *Natural Hazards and Earth System Sciences* 5, 673–677, 2005.

Teotia, S.S., Khattri, K.N., and Roy, P.K. Multifractal analysis of seismicity of Himalayan region. *Current Science* 73, 359–366, 1997.

Teotia, S.S., and Kumar, D. The great Sumatra–Andaman earthquake of 26 December 2004 was predictable even from seismicity data of $m_b \geq 4.5$: A lesson to learn from nature. *Indian Journal of Marine Sciences* 36, 122–127, 2007.

Teotia, S.S., and Kumar, D. Role of multifractal analysis in understanding the preparation zone for large size earthquake in the North-Western Himalaya region. *Nonlinear Processes in Geophysics* 18, 111–118, 2011.

Turcotte, D.L. A model of crustal deformation. *Tectonophysics* 132, 261–269, 1986a.

Turcotte, D.L. Fractals and fragmentation. *Journal of Geophysical Research* 91(B2), 1921–1926, 1986b.

USGS home page, [http://neic.usgs.gov/mneis/epic].

Utsu, T. A method of determining the value of b in a formula log n = a-bM showing the magnitude–frequency relation for earthquakes (in Japanese). *Geophysical Bulletin of Hokkaido University* 13, 99–103, 1965.

Yin, A. Cenozoic tectonic evolution of Himalayan orogen as constrained by along-strike variation of structural geometry, exhumation history, and foreland sedimentation. *Earth-Science Reviews* 76, 1–131, 2006.

Yin, A., Dubey, C.S., Kelty, T.K., Webb, A.A.G., Harrison, T.M., Chou, C.Y., and Célérier, J. Geologic correlation of the Himalayan orogen and Indian craton: Part 2. Structural geology, geochronology, and tectonic evolution of the Eastern Himalaya. *Geological Society of America Bulletin* 122, 360–395, 2010.

7

Geomagnetic Jerks: A Study Using Complex Wavelets

E. Chandrasekhar, Pothana Prasad, and V. G. Gurijala

CONTENTS

Introduction 195
Role of Wavelets and Complex Wavelets in Geomagnetism 199
Data Processing of Global Magnetic Observatory Data 199
Wavelet Analysis 201
    Estimation of Ridge Functions Using Real Wavelets 202
    Complex Wavelet Analysis of GJs 204
Region-Wise Study of GJs 204
    Polar Region 204
    High-Latitude to Mid-Latitude Region 207
    Mid-Latitude to Equatorial Region 209
    Southern Hemisphere Region 209
Discussion 209
Acknowledgments 216
References 216

Introduction

The naturally produced geomagnetic field variations recorded at the earth's surface or at satellite altitudes is of dual origin: external and internal. The external part originates from the current systems generated in the iono-sphere and the distant magnetosphere by the interaction of the solar wind with them. As a result, different types of geomagnetic variations based on the solar intensity having different periodicities are generated. A small portion of the internal part arises due to the eddy currents induced in the conductive layers of the earth by the time-varying external field diffusing within the earth. In addition to induction by external sources, the major contribution for the internal part, known as the earth's permanent magnetic field or dipole field, originates from the earth's liquid outer core. This is a slowly varying field with time. More than 95% of the earth's magnetic field is believed to be due to the electric currents generated within the fluid outer core. An additional contribution of approximately 2% comes from the earth's lithosphere and crust. Thus, the science of geomagnetism opens up a plethora of research topics related to terrestrial and extraterrestrial phenomena, and facilitates the studies of external source current systems, secular variations, and electrical conductivity variations within the earth. Figure 7.1 depicts a schematic of the broad classification of geomagnetic field variations. Figure 7.2 describes the vector representation of the earth's magnetic field at any point in the northern hemisphere of the earth.

Figure 7.3a shows an example of long-period geomagnetic field variations, whose first time-derivative, known as geomagnetic secular variations (Figure 7.3b), owe their origin to the convection currents generated by the dynamo processes actively taking place within the earth's liquid outer core. It has been observed that the geomagnetic data of several decades show some sudden changes in the slope of the secular variations (i.e., second time-derivative of the long-period geomagnetic field), having periods of approximately 1 year. Such features are generally known as secular accelerations or geomagnetic jerks (GJs; Figure 7.3c; Courtillot et al. 1978; Courtillot and Le Mouel 1976, 1984; Backus 1983; Gubbins and Tomlinson 1986; Mandea et al. 2000). Barraclough et al. (1989) described the use of horizontal components of the magnetic field to understand the outer core motions. GJs are easily noticeable in the east–west component (Y or D, Figure 7.2) of the magnetic field, although it is sometimes seen in the north–south component (X or H, Figure 7.2) as well (Bloxham et al. 2002). Well-known global jerks have occurred

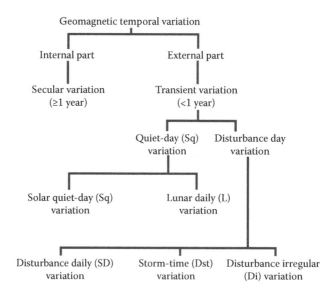

FIGURE 7.1
A schematic showing different types of temporal geomagnetic field variations.

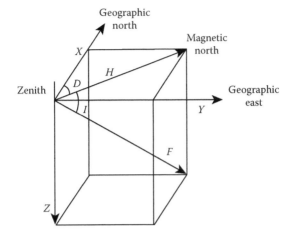

FIGURE 7.2
Elements of the geomagnetic field in northern hemisphere. The declination (*D*) refers to the angle between geographic north (*X*) and geomagnetic north (*H*). The inclination (*I*) refers to the angle between the geomagnetic north and the total field (*F*). *Y* denotes the geographic east component and *Z* refers to the direction of the vertical magnetic field.

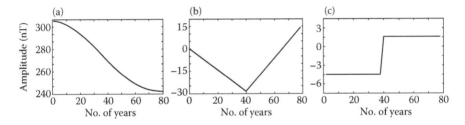

FIGURE 7.3
A schematic for explanation of GJ phenomenon. (a) Example of decadal variations of geomagnetic east–west component, *Y*; (b) secular variations, obtained by taking dY/dt; (c) secular acceleration (d^2Y/dt^2) or GJ, showing a sudden jump at the point of inflection in (a).

during the years 1969 to 1970, 1978 to 1979, and 1991 to 1992 and local jerks during 1901 to 1902, 1913 to 1914, 1925 to 1926, 1932 to 1933, 1942 to 1943, 1949 to 1950, around 1985, and in 1999 to 2000 (Alexandrescu et al. 1995, 1996; De Michelis et al. 1998; De Michelis and Tozzi 2005; Adhikari et al. 2009).

The origin of GJs has been the subject of a lot of debate. Employing a spherical harmonic analysis (SHA) technique, Malin and Hodder (1982) reported that GJs are of internal origin. Later, claiming some loopholes in Malin and Hodder's analysis, Alldredge (1984) argued that GJs are of external origin. However, further studies by Gavoret et al. (1986), Davis and Whaler (1997), Le Huy et al. (1998), Nagao et al. (2002), and Bloxham et al. (2002) confirmed the internal origin of GJs. Gavoret et al. (1986) separated the internal and external field components using geomagnetic indices and found a negligible influence of external sources on GJs. Using the time-dependent flow model of Bloxham and Jackson (1992), Davis and Whaler (1997) studied the outer core fluid flow behavior with respect to the 1969 GJ event and showed that GJs cannot be produced by an instantaneous change in the flow. However, a steady flow with one spherical harmonic flow coefficient changing with time can produce a GJ in the secular variation field at the surface of the earth. Computing the spherical harmonic models for the 1969, 1978, and 1991 jerk events, Le Huy et al. (1998) discussed the characteristics of these successive GJs. Nagao et al. (2002) applied statistical time series modeling to the monthly means of geomagnetic data and discussed the noninfluence of external currents (field-aligned current and ring current) on the occurrence of GJs. Bloxham et al. (2002) explained that the GJs arise due to the combination of a steady flow and a simple time-varying, axisymmetric, equatorially symmetric, toroidal zonal flow of the core fluid. They also opined that the short duration of GJs is due to the differential fluid flow at the surface of the earth's outer core. Therefore, based on the theoretical and observational evidence, it is now well established that GJs are of internal origin.

GJs are believed to occur sometimes globally and sometimes locally,* which has led to the belief of the existence of hemispherical differences in their occurrences (Backus 1983). Differences in their spatiotemporal occurrences have led to studies in the role of lower mantle conductivity as one of the reasons for such a diverse nature of occurrences of GJ events (Acache et al. 1980; Ducruix et al. 1980; Backus 1983; Alexandrescu et al. 1999; Nagao et al. 2003).

In the present study, for the first time, we introduce complex wavelets and demonstrate how they would be useful in understanding the phase characteristics of GJs, which in turn are related to the fluid flow directions in the outer core, which is responsible for the generation of GJs. Using real wavelets, we first discuss the presence or absence of GJs in the global ground magnetic observatory data of 75 years from 74 observatories, by evaluating ridge functions, which signify a log-linear relation between the absolute value of wavelet transform coefficients and the wavelet scale (see Alexandrescu et al. 1995). Next, using complex wavelets, we discuss the phase characteristics of GJs (global or local). We follow a systematic approach by making a groupwise study of global magnetic observatory data corresponding to (*i*) polar regions, (*ii*) high-latitude to mid-latitude regions, (*iii*) mid-latitude to equatorial regions, and (*iv*) southern hemisphere regions. For the first time, we demonstrate that phase information, together with ridge functions, provides an unequivocal confirmation not only about the occurrence of a GJ in the data but also about the changes in the direction of fluid flow at the time of its occurrence. Finally, we discuss the spatiotemporal behavior of GJs in the light of some hemispherical differences observed at the time of their occurrences.

---

* The terms *global* and *local* are purely relative, and as to how global is "global" and how local is "local" are not well defined.

Role of Wavelets and Complex Wavelets in Geomagnetism

Occurrences of GJs in the decadal secular variation data have been understood to be random phenomena and show not only a varied spatiotemporal behavior (sometimes occurring as a global feature and sometimes as a local one) but also a varied time–frequency localization (i.e., irregular occurrences of these high-frequency GJs in time). As a result, conventional signal processing techniques fail to facilitate a better understanding of the overall characteristics of GJs. Wavelet analysis is an effective mathematical tool to achieve the time–frequency localization of GJs (see the Wavelets in Geophysics section in Chapter 1). Such natural events, having diverse space–time and time–frequency characteristics, can be well understood with wavelet schemes (see for example, Alexandrescu et al. 1995; Adhikari et al. 2009). Another interesting characteristic of GJs is that they occur regardless of the direction in the convective motion of the outer core fluid flow (clockwise or anticlockwise). Such fluid flow directions clearly manifest as phase changes in GJs, which can be well understood with the help of complex wavelets. Complex wavelets provide more relevant information on the phase characteristics of the transients present in the signal than real wavelets do (e.g., Unser and Aldroubi 1996; Haddad et al. 2004; Zhou et al. 2009).

Data Processing of Global Magnetic Observatory Data

The geomagnetic east–west component (D or Y), usually recorded in units of arc minutes at any observatory, were first converted into magnetic field units (nanotesla, nT), using

$$D(\text{nT}) = \left[\frac{\pi}{180} \times \frac{H_{\text{abs}}}{60} \right] \times D(\text{min}) \tag{7.1}$$

H_{abs} refers to the absolute value of H component at that magnetic observatory. Similarly, to convert Y variations at any observatory from the units of arc minutes to nanotesla, substitute X_{abs} of that observatory for H_{abs} in the above equation.

Monthly means of D (or Y) variations were used for the present study. They were estimated using the daily mean values, which in turn were calculated with hourly mean values procured from the world data center, WDC-C2 for Geomagnetism, Kyoto University, Japan, and from WDC at Edinburgh. The global magnetic observatory data considered for the present study correspond to the years 1925 to 2000 (Figure 7.4). At places, wherever a whole month's data were missing, they were interpolated using the average of the previous and

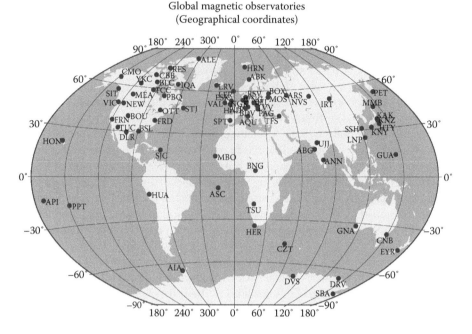

FIGURE 7.4

Global distribution of the 74 geomagnetic observatories, whose data of 75 years (from 1925 to 2000) were used in the present study. The locations of observatories are in geographic coordinates. The three-letter names of the observatories represent the International Association of Geomagnetism and Aeronomy (IAGA) codes. The map is generated by the Generic Mapping Tools software of Wessel and Smith (1995).

next month's values. Also, occasional instrument noise and baseline shifts (that usually result in the form of box-like jumps or step-like jumps) in the data were manually identified and corrected before further analysis. At some instances, when the data of some observatories were continuously missing for a couple of years at one WDC, care was taken to substitute them by obtaining that station's data corresponding to the missing period from another WDC. In studies of the kind being discussed here, such corrections are always necessary and important to determine the correct time of the occurrence of jerks. Otherwise, there arises a great risk of misrepresenting the time of occurrences of GJs. Figure 7.5 shows an example of plots of raw data (Figure 7.5a) and fully processed data (Figure 7.5b) from Lerwick (LER), one of the high-latitude stations (Figure 7.4 and Table 7.1). Next, using the monthly mean values, annual means were calculated to determine the secular variations using the first-order central difference with an interval of 1 year. Here, we would like the reader to note that a library of such fully processed global long-period magnetic observatory data of 75 years from 74 observatories is currently being maintained at the authors' laboratory at Indian Institute of Technology Bombay, India. The fully processed data are available to any user free of cost.

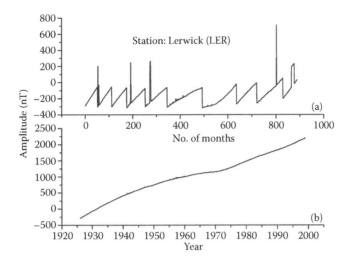

FIGURE 7.5
Raw data (a) and fully processed data (b) of geomagnetic east–west component at one of the high-latitude observatories, LER (see Figure 7.4 and Table 7.1).

Wavelet Analysis

Continuous wavelet transformation (CWT) was performed using real and complex wavelets on the fully processed data of geomagnetic east–west components of approximately 75 years, recorded from 74 global magnetic observatories. Because the occurrences of GJ events are intermittent and sporadic, they can be treated as singularities in the decadal variations of the magnetic field. Accordingly, CWT is ideally suited to detect them. It may be recalled from Chapter 1 that in wavelet transformation, the signal to be transformed is convolved with the mother wavelet and the transformation is computed for different segments of the data at different translations and dilations (see the subsection Wavelets in the Basic Theory and Mathematical Concepts of Wavelets and Fractals in Chapter 1). The wavelet transformation in which the dilations and translations will be continuously varied is called CWT (see the Continuous Wavelet Transformation section in Chapter 1), and the transformation in which they will be varied as power of an integer n (that is, n^j, $j = 1, 2, 3,... k$) is called discrete wavelet transformation (DWT; see the Discrete Wavelet Transformation section in Chapter 1). Generally, in DWT, dyadic scales and shifts are used, in which case $n = 2$. Although both CWT and DWT are linearly translational-invariant (see the Translational Invariance section in Chapter 1), the difference between them is as follows. In CWT, an amount of small time shift (δt) in the wavelet function results in the same amount of shift in the transformed domain. As a result, abrupt changes in the data can be effectively picked up by performing the computations with continuous shifts. On the other hand, in DWT, the wavelet transform coefficients become translational-invariant only if the translations are in dyadic steps [that is, as a power of 2; see Ma and Tang (2001)]. Therefore, CWT, in general, is best suited for such studies in which abrupt jumps, discontinuities, spikes, singularities, etc., in the data are of interest for analysis.

CWT with real wavelets was first performed to ensure the presence or absence of GJs in the decadal variations of magnetic data. Next, the CWT with complex wavelets was done to study the phase characteristics of the identified GJs. The presence of a GJ in the data can be identified by examining the linear relation between the logarithm of the absolute value of wavelet transform coefficients and the logarithm of the wavelet scale.

Estimation of Ridge Functions Using Real Wavelets

Because the occurrences of GJ events are causal* in nature, they can be treated as independent events (singularities) and accordingly, each event can be analyzed separately. Furthermore, because they manifest the second time-derivative of the geomagnetic field, these singularities can be represented as some αth derivative of the signal (α being the regularity of the singularity). So, the time-varying magnetic field, say, $f(t)$ can be expressed as (Alexandrescu et al. 1995)

$$f(t) = A(t + \delta t)^\alpha + h(t) + n(t) \tag{7.1}$$

where $h(t)$ is the long-period harmonic component and $n(t)$ is the noise term, as a function of time. The factor $A(t + \delta t)$ with the condition

$$(t + \delta t)^\alpha = \begin{cases} 0 & t \leq \delta t \\ (t + \delta t)^\alpha & t > \delta t \end{cases} \tag{7.2}$$

defines the presence of a singularity in the signal centered around δt, having amplitude A. Because the singularity is included in $f(t)$, the wavelet transformation can detect it and can also tell about the time of its occurrence. Performing wavelet transformation on Equation 7.1 and doing some subsequent simple mathematical calculations (see Alexandrescu et al. 1995), we obtain ridge functions that represent a linear regression between the logarithm of the absolute value of wavelet transform coefficients along the lines of maxima, $\log_2(|CWT_{\tau,s}|)$, and the logarithm of scale, $\log_2(s)$. Mathematically, it is expressed as

---

* A time signal $f(t)$ is said to be causal if it is zero before the time of its occurrence. Mathematically,
  it is represented as $\begin{aligned} f(t) &= 0 & t &< 0 \\ &= f(t) & t &\geq 0. \end{aligned}$

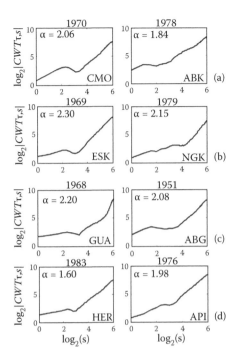

FIGURE 7.6
Ridge function plots of a few stations, each corresponding to polar region (a), high-latitude to mid-latitude region (b), mid-latitude to equatorial region (c), and southern hemisphere (d) stations. The regularity value or the slope, α, is computed by fitting a linear fit to the data beyond the scale $2^{3.5}$.

$$\log_2(|CWT_{\tau,s}|) = \alpha \log_2(s) + \text{constant} \qquad (7.3)$$

where $s > 0$ is the dilation (scale) parameter. Because GJs represent the second time-derivative of the magnetic field, the computed value of α (the slope of Equation 7.3) should be equal to or close to 2. For full mathematical details of derivation of Equation 7.3, the reader is referred to Alexandrescu et al. (1995) and the references therein. Among a variety of real wavelets tested, we have found the Gaus3 wavelet as the optimum one for identifying GJs. Figure 7.6 shows the ridge function plots obtained by using Gaus3 wavelet corresponding to GJ events observed at a few stations in the polar region (Figure 7.6a), high-latitude to mid-latitude region (Figure 7.6b), mid-latitude to equatorial region (Figure 7.6c), and southern hemisphere stations (Figure 7.6d). The α value or the slope is computed by fitting a linear fit to the data beyond the scale $2^{3.5}$.

Complex Wavelet Analysis of GJs

The complex continuous wavelet transform (CCWT), like the real wavelet transform, expands a signal in terms of a set of complex wavelet basis functions, $\psi^*_{\tau,s}$, which are dilated (scaled) and translated (shifted) versions of the complex wavelet. The real and imaginary parts of complex mother wavelet, $\psi^*(t)[=\psi_r(t)+i\psi_i(t)]$, individually satisfy all the conditions of the real-valued mother wavelets. The CCWT coefficients, $CCWT_{\tau,s}$, are estimated by

$$CCWT_{\tau,s} = \frac{1}{\sqrt{s}} \int f(t)\psi^*\left(\frac{t-\tau}{s}\right)dt. \tag{7.4}$$

The complex Gaussian wavelet family is derived from the derivative of Gabor wavelet, given as

$$\psi^*(t) = C\frac{d^n}{dt^n}G(t) \tag{7.5}$$

where $G(t) = e^{-(t^2+it)}$ defines the Gabor wavelet, n denotes the order of the derivative, and C is the constant. We have used a variety of complex wavelets, and based on the resolution of the phase information that they had offered, we have identified the complex Gaussian wavelet 1 (Cgau 1, i.e., $n = 1$ in Equation 7.5) as the most suitable one for the present study. From Equation 7.5, it can be easily shown that the real and imaginary parts of the complex Gaussian wavelet form a Hilbert transform pair for all n (Selesnick 2001). As a result, the wavelet transform coefficients, $|CCWT_{\tau,s}|$ become translational invariant (Fernandes 2001; Kingsbury 2001).

Region-Wise Study of GJs

Polar Region

We have analyzed approximately 70 years of data from 11 observatories from this region (Table 7.1). The 1969 to 1970 global jerk event is well resolved in the ABK, ALE, BLC, CMO, LER, and LRV observatories. At all these observatories (except at CMO), the 1969 to 1970 jerk event in the secular variation data is characterized by a convex-shaped curve (slope changing from negative to positive), whereas at CMO, it is characterized by a concave shape (slope changing from positive to negative). An example of phase scalogram plots together with secular variation plots and the ridge function plots related to the 1969 to 1970 event corresponding to

TABLE 7.1

Geographic Coordinates and IAGA Codes of Magnetic Observatories Shown in Figure 7.4

Sl. No.	IAGA Code	Station Name	Latitude (°N)	Longitude (°E)
Observatories Corresponding to the Polar Region				
1.	ABK	Abisko	68.358	18.823
2.	ALE	Alert	82.497	297.647
3.	BLC	Baker Lake	64.333	263.967
4.	CBB	Cambridge Bay	69.123	254.969
5.	CMO	College	64.87	212.14
6.	HRN	Hornsund	77.0	15.55
7.	IQA	Iqaluit	63.75	291.482
8.	LER	Lerwick	60.133	358.817
9.	LRV	Leirvogur	64.183	338.3
10.	RES	Resolute Bay	74.69	265.105
11.	YKC	Yellowknife	62.482	245.518
Observatories Corresponding to the High-Latitude to Mid-Latitude Regions				
12.	AQU	L'Aquila	42.383	13.317
13.	ARS	Arti	56.433	58.567
14.	BDV	Budkov	49.08	14.015
15.	BEL	Belsk	51.837	20.792
16.	BOU	Boulder	40.14	254.767
17.	BOX	Borok	58.033	38.972
18.	BSL	Stennis Space Centre	30.35	270.36
19.	ESK	Eskdalemuir	55.317	356.8
20.	FCC	Fort Churchill	58.786	265.912
21.	FRD	Fredericksburg	38.21	282.633
22.	FRN	Fresno	37.09	240.283
23.	FUR	Furstenfeldbruck	48.17	11.28
24.	HAD	Hartland	51	355.517
25.	HTY	Hatizyo	33.073	139.825
26.	IRT	Patrony	52.167	104.45
27.	KAK	Kakioka	36.232	140.186
28.	KNY	Kanoya	31.424	130.88
29.	KNZ	Kanozan	35.256	139.956
30.	LVV	Lvov	49.9	23.75
31.	MEA	Meanook	54.615	246.653
32.	MMB	Memambetsu	43.91	144.189
33.	MOS	Krasnaya Pakhra	55.467	37.317
34.	NEW	Newport	48.27	242.883
35.	NGK	Niemegk	52.072	12.675
36.	NVS	Novosibirsk	54.85	83.23
37.	OTT	Ottawa	45.403	284.448

(*continued*)

TABLE 7.1 (Continued)

Geographic Coordinates and IAGA Codes of Magnetic Observatories Shown in Figure 7.4

Sl. No.	IAGA Code	Station Name	Latitude (°N)	Longitude (°E)
38.	PAG	Panagjurishte	47.485	24.177
39.	PBQ	Poste-de-la-Baleine	55.277	282.255
40.	PET	Paratunka	52.971	158.248
41.	RSV	Rude Skov	55.85	12.45
42.	SIT	Sitka	57.067	224.67
43.	SPT	San Pablo-Toledo	39.547	355.651
44.	SSH	Sheshan (Zo-Se)	31.097	121.187
45.	STJ	Saint Johns	47.595	307.323
46.	TFS	Dusheti (Tblisi)	42.092	44.705
47.	TUC	Tucson	32.17	249.27
48.	VAL	Valentia	51.933	349.75
49.	VIC	Victoria	48.517	236.583
50.	WNG	Wingst	53.743	9.073
Observatories Corresponding to the Mid-Latitude to Equatorial Regions				
51.	ABG	Alibag	18.638	72.872
52.	ANN	Annamalainagar	11.367	79.683
53.	BNG	Bangui	4.333	18.567
54.	DLR	Del Rio	29.49	259.083
55.	GUA	Guam	13.59	144.87
56.	LNP	Lunping	25.0	121.167
57.	HON	Honolulu	21.32	202
58.	MBO	MBour	14.384	343.033
59.	SJC	Cayey, PR	18.11	293.85
60.	UJJ	Ujjain	23.18	75.78
Observatories Corresponding to the Southern Hemisphere				
61.	AIA	Faraday Islands	−65.245	295.742
62.	API	Apia	−13.815	188.219
63.	ASC	Ascension Island	−7.949	345.624
64.	CNB	Canberra	−35.32	149.36
65.	CZT	Port Alfred	−46.431	51.867
66.	DRV	Dumont d'Urville	−66.667	140.007
67.	DVS	Davis	−68.583	77.967
68.	EYR	Eyrewell	−43.422	172.355
69.	GNA	Gnangara	−31.8	116
70.	HER	Hermanus	−34.425	19.225
71.	HUA	Huancayo	−12.05	284.67
72.	PPT	Pamatai	−17.566	210.416
73.	SBA	Scott Base	−77.85	166.763
74.	TSU	Tsumeb	−19.202	17.584

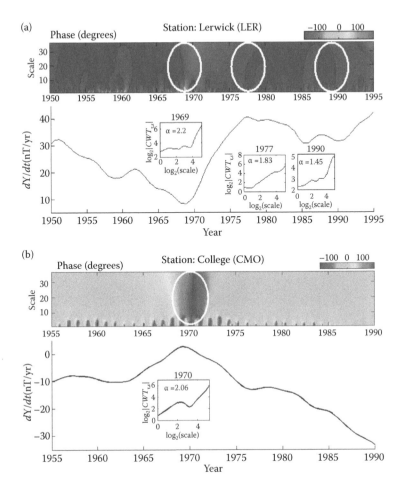

FIGURE 7.7
(See color insert.) Phase scalogram plots together with secular variation plots and the ridge function plots related to the 1969–1970 event corresponding to (a) LER and (b) CMO observatories of polar region.

the LER and CMO observatories are shown in Figure 7.7. Although LER (Figure 7.7a) shows all three global jerk events, CMO (Figure 7.7b) clearly shows only the well-resolved 1969 to 1970 event. However, a study of the phase scalogram plots of all the observatories of this region reveals that the occurrences of all the above three global jerk events were sporadic and not seen at most of the observatories in this region.

High-Latitude to Mid-Latitude Region

We have analyzed approximately 75 years of data from as many as 39 observatories from this region (Table 7.1), since most observatories are located in

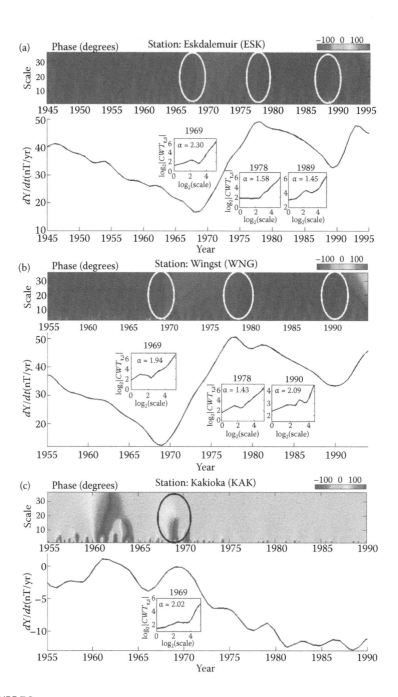

FIGURE 7.8
(See color insert.) Phase scalogram plots together with secular variation plots and the ridge
function plots related to the global jerk events corresponding to (a) ESK, (b) WNG, and (c) KAK
observatories of the high-latitude to mid-latitude regions.

this region. Figure 7.8 shows the secular variation, phase scalogram plots, and the associated ridge functions corresponding to the three global jerk events of ESK (Figure 7.8a), WNG (Figure 7.8b), and KAK (Figure 7.8c) observatories. At KAK and other Japanese observatories, except for the 1969 event, the other two events could not be well resolved (Figure 7.8c). In this region, the three global jerk events—1969 to 1970, 1978 to 1979, and 1991 to 1992—do not show any differences in the phases corresponding to the same events observed at other observatories, and their corresponding ridge functions also showed a near log-linear nature at these events.

Mid-Latitude to Equatorial Region

We have analyzed about 70 years of data from 10 observatories from this region (Table 7.1). Figure 7.9 shows the secular variation, phase scalogram plots, and the associated ridge functions corresponding to the global as well as local jerk events, as observed at the ABG (Figure 7.9a), BNG (Figure 7.9b), and MBO (Figure 7.9c) observatories. ABG does not show any global events, but shows two clear local jerk events, which occurred during 1950 to 1951 and 1984 to 1985 (Figure 7.9a). In the secular variation plot of BNG (Figure 7.9b), the well-known global GJ event of 1969 showing a concave shape is immediately followed by a well-defined convex shape at 1971. It is intriguing to observe two well defined jerk events within a gap of two years. A similar observation of change of phase is also clearly seen in 1983 (phase changing from positive to negative) and 1990 (phase changing from negative to positive) in a gap of seven years. At MBO (Figure 7.9c), well-defined GJ events showing clear phase changes are seen at the years 1967 and 1982. At almost all observatories in this region, the ridge functions corresponding to global jerk events showed a near log-linear nature and also there are no appreciable phase differences corresponding to the respective events observed at other observatories.

Southern Hemisphere Region

Data from only 14 observatories were available for analysis corresponding to this region (Table 7.1). Figure 7.10 shows the secular variation, phase scalogram plots, and the associated ridge functions corresponding to the global as well as local jerk events, as observed at the HER (Figure 7.10a), PPT (Figure 7.10b), and API (Figure 7.10c) observatories.

Discussion

The GJs that arise due to the differential fluid flow in the outer core are clearly identified by the changes in the slopes of secular variations. They show up as

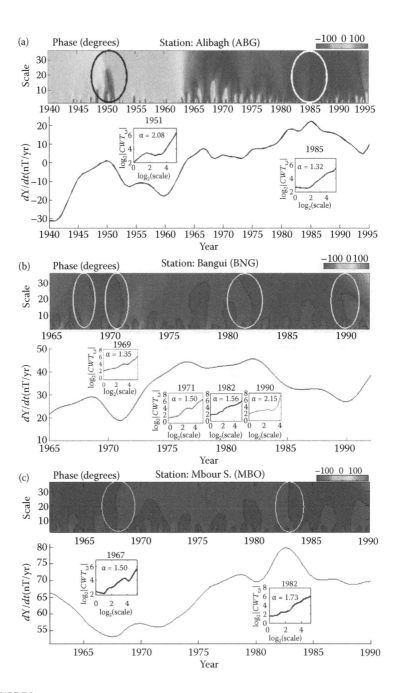

FIGURE 7.9
(See color insert.) Phase scalogram plots together with secular variation plots and the ridge function plots related to the global and local jerk events corresponding to (a) ABG, (b) BNG, and (c) MBO observatories of the mid-latitude to equatorial regions.

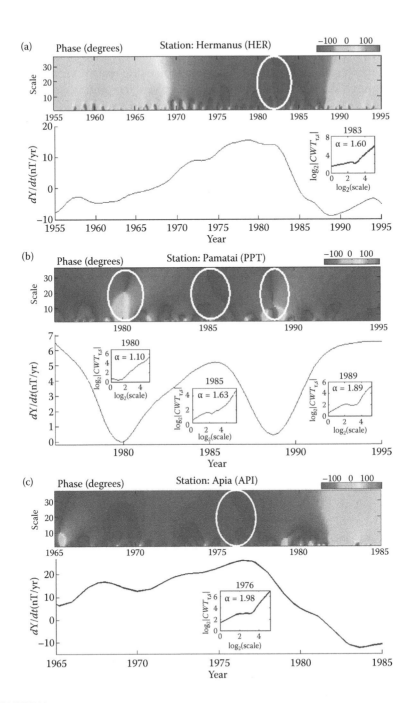

FIGURE 7.10
(See color insert.) Phase scalogram plots together with secular variation plots and the ridge function plots related to the local jerk events corresponding to (a) HER, (b) PPT, and (c) API observatories of the southern hemisphere region.

convex-shaped or concave-shaped curves in the secular variation data, whose extremum (local maximum or minimum) signifies the time of occurrence of jerks. In this study, an analysis of GJs was done using real and complex wavelets. Although the former were used to first ensure the presence or absence of GJs in the data, the latter were used to understand the phase characteristics of the identified GJs. As explained earlier, the linear regression between the logarithm of wavelet transform coefficients and the logarithm of wavelet scale (i.e., the ridge function) confirms the presence or absence of GJs in the data. Figure 7.6 shows the ridge function plots depicting the occurrence of some global as well as some local GJs at some select observatories corresponding to different regions on the globe. The ridge functions associated with jerk events are linear at scales $>2^{3.5}$ (~12). Because the monthly mean values were used in the analysis, perhaps this number is related to the 1-year period associated with jerk events. The portions of the ridge function curves at scales $<2^{3.5}$, which are not linear, are associated with extraterrestrial noise (Alexandrescu et al. 1995) affecting the data. The scale values in the range $2^{3.5}$ to 2^4 mark the transition between the part of the curve dominated by external noise and the part dominated by GJs. If the influence of external noise is high in the data, then the identification of time of occurrence of the jerk event will be less accurate. However, at some observatories, for a couple of ridge functions, the linear nature of the curves starts at scale values less than $2^{3.5}$ [see, for example, the 1978 event at ESK (Figure 7.8a) and the 1969 and 1978 events at WNG (Figure 7.8b)]. This possibly suggests that the wavelet may be averaging out the external noise during the CWT operation. At some stations, the linear nature of the curves starting at higher scales (i.e., $s > 2^4$) is less perfect, in the sense that some kinks are observed in the linear portion of the curves. It is not clear, if a strong long-period external noise in the data is responsible for this. Furthermore, it is also important to note that the α values estimated from Equation 7.3 are not the intrinsic property of the jerk alone, but also depend on the wavelet used (Adhikari et al. 2009).

Haddad et al. (2004) show that the phase scalogram plots corresponding to the complex wavelet transform of synthetic data suggest a clear phase change from $-\pi$ to $+\pi$ (corresponding to convexity) and $+\pi$ to $-\pi$ (corresponding to concavity) at different discontinuities present in the data. In case of GJ phenomena, such phase changes imply changes in the direction of large-scale convective motions on the surface of the outer core. A careful observation of phase scalogram plots together with respective secular variation data and the associated ridge functions corresponding to the global and local jerks shown in Figures 7.7 through 7.10 provide some useful clues for better understanding the nature of the fluid flow at the time of occurrence of different GJs.

The European observatories show a clear signature of the occurrence of global jerks (Figure 7.8a and b). The near log-linear nature of the ridge functions corresponding to these jerks at most European observatories also supports this. Furthermore, there is also a clear change in the phase of these discontinuities from $-\pi$ to $+\pi$ or $+\pi$ to $-\pi$ at the time of the occurrence of GJs. A careful examination of Figures 7.7a and 7.8b shows a clear double peak in the secular variation data around the time of the 1978 to 1979 jerk event at a high-latitude station (LER) and at a European station (WNG), respectively. Around the same time, another European observatory, Niemegk (NGK), also observed such a feature (not shown here). This phenomenon is understood to be a manifestation of time-varying outer core fluid flow (Bloxham et al. 2002). At most of the observatories, while the 1969–1970 global event shows a clear phase change from $-\pi$ to $+\pi$, the corresponding phase changes for the same events at CMO (Figure 7.7b) and KAK (Figure 7.8c), are from $+\pi$ to $-\pi$. It is quite intriguing to know what could have caused such an oppositely directed secular acceleration at these two northern hemisphere stations? Secondly, it is also interesting and important to understand, what drives these secular accelerations to reverse their directions within a span of about two decades (the three jerks: 1969, 1978, and 1991 occurred in about two decades), whereas such signatures are not seen at other observatories located in the same hemisphere. Perhaps knowledge on the spatial scale lengths of the outer core convective motions would be quite helpful. Figure 7.11 shows a summary of worldwide distribution of phase changes associated with the three global jerk events of 1969 to 1970 (Figure 7.11a), 1978 to 1979 (Figure 7.11b), and 1991 to 1992 (Figure 7.11c). They show a clear understanding of the nature of outer core fluid motions associated with these jerk events. Figures 7.11a to 7.11c also summarize the hemispherical differences not only in the occurrences of global GJ events but also in their phase characteristics. The poor and sparse distribution of available data in the southern hemisphere also suggests an urgent need to fill such a wide gap in establishing more observatories, so as to have an improved understanding of the GJ phenomena and their associated outer core fluid motions in a broader perspective.

One can also easily observe that the secular variation plots corresponding to most of the observatories show changes in slopes at several years other than the global or local jerk years (Figures 7.7–7.10). The corresponding phase transitions at these years are also notable, particularly only at smaller scales. However, their associated ridge functions do not show the log-linear nature of the curves at higher scales at these years. Perhaps the acceleration in the outer core fluid was not strong enough to drive a GJ.

In addition to the global jerks, the characteristic features seen in secular variations during the years 1951 at ABG (Figure 7.9a), 1983 at HER (Figure 7.10a), 1980 thru 1989 at PPT (Figure 7.10b) and 1976 at API (Figure 7.10c) can be classified as local jerks. They show clear phase changes observed in the complex wavelet phase scalogram plots together with good log-linear nature of ridge functions at these years. This clearly warrants that they be identified as local jerks. Figure 7.9a shows clear change in slope of secular variation data around 1951 at ABG. However, the resolution of the change in phase corresponding to this jerk event in the phase scalogram plot is somewhat poor. We attribute this to the noise associated with this coastal station data.

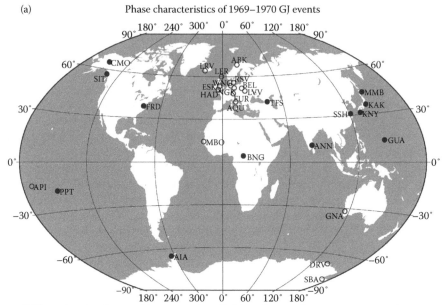

(a) Phase characteristics of 1969–1970 GJ events

● Phase changing from positive to negative   ○ Phase changing from negative to positive

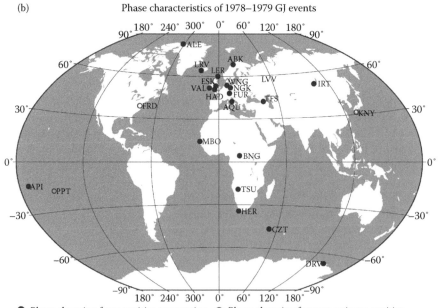

(b) Phase characteristics of 1978–1979 GJ events

● Phase changing from positive to negative   ○ Phase changing from negative to positive

FIGURE 7.11
A summary of worldwide distribution of phase changes associated with the three global jerk events of 1969 to 1970 (a), 1978 to 1979 (b), and 1991 to 1992 (c).

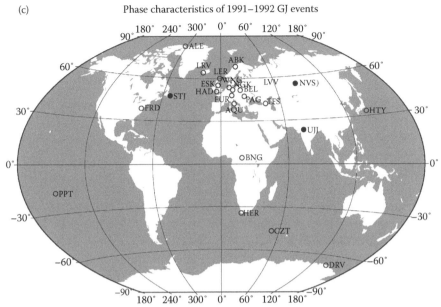

(c)   Phase characteristics of 1991–1992 GJ events

● Phase changing from positive to negative   ○ Phase changing from negative to positive

FIGURE 7.11 (Continued)
A summary of worldwide distribution of phase changes associated with the three global jerk
events of 1969 to 1970 (a), 1978 to 1979 (b), and 1991 to 1992 (c).

Such a poor resolution in phase change is also seen at another coastal station, KAK (see Figure 7.8c corresponding to the 1969 jerk event). Further, the
shape of the secular variation curve associated with the 1951 local jerk at
ABG (Figure 7.9a) and the 1969 global jerk at KAK (Figure 7.8c) is rather broad
and, accordingly, the phase change is not sharp at the occurrence of this jerk
but gradual.

A careful observation of phase information of all the global GJ events summarized in Figure 7.11 deduced from the complex wavelet transformation,
has provided unequivocal evidence that there exists some hemispherical differences in the occurences of global jerks.

We conclude that complex wavelets have helped to understand the nature
of outer core fluid motions associated with global as well as local GJs. The
hemispherical differences and the delays observed in the time of occurrence
of GJ phenomena can be associated with either poor distribution of observatories in the southern hemisphere or a mantle filtering effect as advocated by
Backus (1983). Alexandrescu et al. (1999) and Nagao et al. (2003) also relate
the delays in observing the jerk features to variations in mantle conductivity.
However, this enigma can be solved if more ground magnetic observatories
are established in the southern hemisphere and continuously monitored for
many years. It is also clearly evident from this study, further to ridge function

plots of GJ events, phase scalogram plots also help in improving the overall understanding of the morphology of global and local GJ events.

The following illustrate some unanswered questions on this topic, in which wavelets can play a significant role in providing a useful and complementary class of information in their effective understanding.

(*i*)   Why are GJ events clearly seen only in the east–west component of the magnetic field?

(*ii*) What drives the convection in the outer core to be global at some instances and local at other times?

(*iii*) What makes the hemispherical differences in the occurrences of GJs? Are these time delays solely attributed to lower mantle conductivity variations alone or are any other dynamics in place? Will knowledge of spatial scale lengths of the outer core convective motions be helpful in understanding this?

(*iv*) What are the implications of GJs on lower mantle conductivity and vice versa? A further detailed study on better understanding of the morphology of the GJ phenomena, which would provide some answers to the above questions, is important and necessary.

(*v*)   Finally, are GJs predictable?

Acknowledgments

The authors thank ISRO-RESPOND, Department of Space, Government of India, for providing financial support in the form of a sponsored research project. They express their sincere thanks to Dr. Nandini Nagarajan and an anonymous referee for their constructive criticism and meticulous reviews, which have improved the quality of the chapter. Dr. K.V. Praveen Kumar, V. Eswara Rao, K.V. Gopalakrishnan, S. Adhikari and K. Sravanti are thanked for their help at various stages of data processing.

References

Acache, J., Courtillot, V., Ducruix, J., and Le-Mouël, J.L. 1980. The late 1960's secular variation impulse: Further constraints on deep mantle conductivity. *Physics of Earth Planetary Interriors* 23, 72–75.

Adhikari, S., Chandrasekhar, E., Rao, V.E., and Pandey, V.K. 2009. On the wavelet analysis of geomagnetic jerks of Alibag magnetic observatory data, India. *Proceedings of the XIII IAGA Workshop on Geomagnetic Observatory Instruments, Data Acquisition and Processing*, edited by Jeffery J. Love, USGS open-file report 2009-1226, 14–23.

Alexandrescu, M., Gilbert, D., Hulot, G., Le Mouël, J.-L., and Saracco, G. 1995. Detection of geomagnetic jerks using wavelet analysis. *Journal of Geophysical Research* 100, 12557–12572.

Alexandrescu, M., Gilbert, D., Hulot, G., Le Mouël, J.-L., and Saracco, G. 1996. Worldwide wavelet analysis of geomagnetic jerks. *Journal of Geophysical Research* 101(B10), 21975–21994.

Alexandrescu, M., Gilbert, D., Le Mouël, J.-L., Hulot, G., and Saracco, G. 1999. An estimate of average lower mantle conductivity by wavelet analysis of geomagnetic jerks. *Journal of Geophysical Research* 104, 17735–17746.

Alldredge, L.R. 1984. A discussion of impulses and jerks in the geomagnetic field. *Journal of Geophysical Research* 89, 4403–4412.

Backus, G.E. 1983. Application of mantle filter theory to the magnetic jerk of 1969. *Geophysical Journal of the Royal Astronomical Society* 74, 713–746.

Barraclough, D., Gubbins, D., and Kerridge, D. 1989. On the use of horizontal components of magnetic field in determining core motions. *Geophysical Journal International* 98, 293–299.

Bloxham, J., and Jackson, A. 1992. Time-dependent mapping of the magnetic field at the core-mantle boundary. *Journal of Geophysical Research* 97(B13), 19537–19563. doi: 10.1029/92JB01591.

Bloxham, J., Zatman, S., and Dumberry, M. 2002. The origin of geomagnetic jerks. *Nature* 420(6911), 65–68.

Courtillot, V., and Le-Mouël, J.-L. 1984. Geomagnetic secular variation impulses. *Nature* 311, 709–716.

Courtillot, V. and Le Mouël, J.-L. 1976. On the long-period variations of the Earth's magnetic field from 2 months to 20 years. *Journal of Geophysical Research* 81, 2941–2950.

Courtillot, V., Ducruix, J., and Le Mouël, J.-L. 1978. Sur une accélérationrécente de la variation séculaire du champ magnétique terrestre. *Compestes Rendus de l'Académie des Sciences* 287, Série D, 1095–1098.

Davis, R.G., and Whaler, K. 1997. The 1969 geomagnetic impulse and spin-up of the Earth's liquid core. *Physics of the Earth and Planetary Interiors* 103, 181–194.

De Michelis, P., Cafarella, L., and Meloni, A. 1998. Worldwide character of the 1991 geomagnetic jerk. *Earth and Planetary Science Letters* 25(3), 377–380.

De Michelis, P., and Tozzi, R. 2005. A Local Intermittency Measure (LIM) approach to the detection of geomagnetic jerks. *Earth Planetary Science Letters* 235, 261–272.

Ducruix, J., Courtillot, V., and Le-Mouel, J.L. 1980. The late 1960's secular variation impulse, the eleven year magnetic variation and the electrical conductivity of the deep mantle. *Geophysical Journal of the Royal Astronomical Society* 61, 73–94.

Fernandes, F.C. 2001. Directional, Shift-Insensitive, Complex Wavelet Transforms with Controllable Redundancy. PhD Thesis, Rice University.

Gavoret, J., Gilbert, D., Menvielle, M., and Le Mouël, J.-L. 1986. Long-term variations of the external and internal components of the Earth's magnetic field. *Journal Geophysical Research* 91, 4787–4796.

Gubbins, D., and Tomlinson, L. 1986, Secular variation from monthly means from Apia and Amberley magnetic observatories. *Geophysical Journal of the Royal Astronomical Society* 86, 603–616.

Haddad, S.A.P., Karel, J.M.H., Peeters, R.L.M., Westra, R.L., Serdijn, W.A. 2004. Complex wavelet transform for analog signal processing. *Proceedings ProRISC, Veldhoven, November, 2004.*

Kingsbury, N.G. 2001. Complex wavelets for shift-invariant analysis and filtering of signals. *Journal of Applied and Computational Harmonic Analysis* 10(3), 234–253.

Le Huy, M., Alexandrescu, M., Hulot, G., and Le Mouël, J.-L. 1998. On the characteristics of successive geomagnetic jerks. *Earth, Planets and Space* 50, 723–732.

Ma, K., and Tang, X. 2001. Translation-invariant face feature estimation using discrete wavelet transform, In *Wavelet Analysis and its Applications*, Y.Y. Tang et al. (Eds.), LNCS 2251, 200–210. Springer-Verlag, Berlin, Heidelberg.

Malin, S.R.C., and Hodder, B.M. 1982. Was the 1970 geomagnetic jerk of internal or external origin? *Nature* 296, 726–728.

Mandea, M., Bellanger, E., and Le Mouël, J.-L. 2000. A geomagnetic jerk for the end of the 20th century? *Earth Planetary Science Letters*, 183, 369–373.

Nagao, H., Iyemori, T., Higuchi, T., Nakano, S., and Araki, T. 2002. Local time features of geomagnetic jerks. *Earth, Planets and Space* 54, 119–131.

Nagao, H., Iyemori, T., Higuchi, T., and Araki, T. 2003. Lower mantle conductivity anomalies estimated from geomagnetic jerks. *Journal of Geophysical Research* 108(B5), 2254.

Selesnick, I.W. 2001. The design of Hilbert transform pairs of wavelet bases via the flat delay filter. *Acoustics, Speech, and Signal Processing* 6, 3673–3676.

Unser, M., and Aldroubi, A. 1996. A review of wavelets in biomedical applications. *Proceedings of the IEEE* 84(4), 626–638.

Wessel, P., and Smith, W.H.F. 1995. New version of the generic mapping tools released. *EOS, Transactions of the American Geophysical Union* 76, 329.

Zhou, G.R., Wang, W.J., and Sun, S.X. 2009. Phase properties of complex wavelet coefficients. *Fourth International Conference on Innovative Computing, Information and Control. 978-0-7695-3873-0/09, IEEE.*

8

Application of Wavelet Transforms to Paleomonsoon Data from Speleothems

M. G. Yadava, Y. Bhattacharya, and R. Ramesh

CONTENTS

Introduction 219
Multitaper Analysis 220
Extracting the Low Power, Hidden Cycles during Wavelet Analysis 222
Conclusions 227
Acknowledgments 227
References 227

Introduction

We use stable oxygen isotope variations in cave calcites [$CaCO_3$ stalactites and stalagmites, collectively known as speleothems; see Yadava and Ramesh (1999) for the detailed methodology] to reconstruct past monsoon variations. Caves are among the longest-lived components of landscapes, surviving for millions of years. The most preferred archive for paleoclimatic reconstruction after tree rings and corals are speleothems. Stalagmite, a column-shaped speleothem forming on the cave floor, is like a stone tree growing for many thousands of years. Having the cover of overlying bedrock and soils, these have natural protection against erosion and weathering. Calcite stalagmites, stalactites, and flowstones are useful for environmental reconstruction. For dating speleothems, both radiocarbon and U–Th methods are applied. However, these methods have certain limitations: for example, in any speleothem, the varying fraction of dead carbon derived from the bedrock may result in radiocarbon ages that may not show increasing age for older layers. Also, ^{14}C method cannot be applied beyond approximately 40 to 50 ka when radiocarbon detection is not possible. High fine-dust (detrital) content in the seepage water having dissolved ionic species may produce U–Th ages with large uncertainty as well as age reversals. Thus far, five Indian caves have been explored for generating terrestrial monsoon history of the Indian subcontinent. Here, we use the data from one such cave to study the application of wavelets.

During speleothem formation, stable isotope ratio of oxygen ($^{18}O/^{16}O$) in the HCO_3^- ions in the dripping water are affected by the cave environment. This ratio is preserved in the $CaCO_3$ laminae and is used to decipher the past cave environment. The ratios are measured relative to a laboratory standard on a mass spectrometer and expressed as $\delta^{18}O$, where

$$\delta^{18}O = \left[\left(\left(^{18}O/^{16}O\right)_{sample}/\left(^{18}O/^{16}O\right)_{standard}\right) - 1\right] \times 1000 \quad (\delta \text{ in permil}).$$

In the tropics, $\delta^{18}O$ of rainwater is inversely correlated with amount of rainfall, termed as the amount effect, and hence, speleothem $\delta^{18}O$ may act as a proxy for the amount of rainfall, that is, the strength of the Indian summer monsoon.

Akalagavi (Uttar Kannada district, Karnataka) is a small cave located near the coastline of a peninsular tropical dense forest. An active stalagmite collected from this cave had annual laminations that were used to set up chronology (331 years, from CE 1666 to 1996). In the period covered by the instrumental record, depleted and enriched ^{18}O spikes in the stalagmite are correlatable, respectively, with the flood and drought years in the instrumental rainfall record, within a few years of uncertainty. Moreover, during 1850 to 1920 CE, when instrumental data shows several drought events, clustering of wavelet power associated with a 21-year sunspot cycle is detectable in the $\delta^{18}O$ data (Yadava and Ramesh 2007). Such correlations are indirect robust proofs that speleothem $\delta^{18}O$ is responding to the monsoon variations, and temporal variations in the solar control are observable.

Multitaper Analysis

The variance of a periodogram estimate is the square of its expectation value at a certain frequency, that is, the standard deviation of a periodogram is independent of the value of the finite number of sample data (N). Using a longer sample reduces the bias of the periodogram, that is, it reduces the leakage from one frequency to other frequencies. To reduce the side lobes of spectral leakage, data is windowed before estimating the periodogram. For a rectangular window (or a window with sharp edges), the side lobes can easily lead to spurious frequencies showing up in the spectrum. Blackman and Tukey (1958) introduced a way around this by convolving the data segment with a suitable window function (taper) that smoothly attenuates the amplitude of the input signal to zero at the edges, with the objective of managing frequency leakage and (ideally) minimizing the number of spurious frequencies showing up in the spectrum (modified periodogram). Windowing also effectively discards data toward the segment edges, and there is a loss of spectral resolution due to widening of the main lobe.

Thomson's multitaper method provides a way to determine power spectral density whilst minimizing both variance and the loss of information arising from the use of a single window taper (Thomson 1982). The resolution is not degraded much by using a small set of tapers, theoretically reducing the variance of the spectrum with the number of tapers. In general, a good taper maximizes the spectral energy (relative to the total energy) in a frequency band [–W,W] of width 2W. Discrete prolate spheroidal sequences are often used as tapers, which effectively minimizes leakage outside a frequency with half the bandwidth. When choosing the number of tapers, important characteristics of the data are the length of the data set and the time resolution, as well as the preferred frequency resolution and degrees of freedom (*df*) in the power spectrum. The frequency resolution increases with greater number of tapers, although too many tapers in certain cases tend to increase the leakage.

In the multitaper method, the statistical information discarded by one taper is partially recovered by the others, optimizing resistance to spectral leakage. The multitaper spectrum is constructed by a weighted sum of these single tapered periodograms. The weighting function is defined to generate a smooth estimate with less variance than single-taper methods and, at the same time, to have reduced bias from spectral leakage. In the present study, multitaper analysis was performed on a 331-year data set of proxy monsoon data (Figure 8.1).

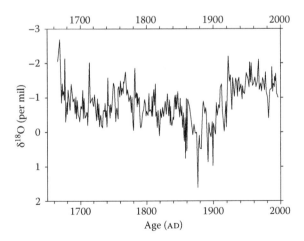

FIGURE 8.1
Raw oxygen stable isotope ratio data ($\delta^{18}O$) produced from a cave carbonate deposit (Yadava et al. 2004) covering a 331-year time period (CE 1666–1996).

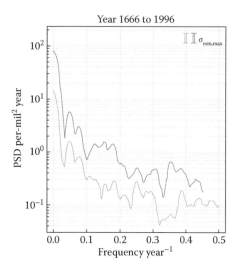

FIGURE 8.2

Multitaper analysis of data in Figure 8.1 shows that both carbon isotopes (black) and hydrogen isotopes (gray) have similar variations and a strong significant power in the 22-year solar cycle (second peak from left). Error bars are shown on the top right corner.

Figure 8.2 shows the multitaper analysis of Akalagavi oxygen and carbon isotope data. Interestingly, both show similar variations and a strong 22-year periodicity.

Extracting the Low Power, Hidden Cycles during Wavelet Analysis

Standard wavelet codes are usually available for novice users. Mostly, these are in the form of executable files or precisely defined codes and, therefore, to incorporate any change in the tool pack, one requires a thorough knowledge of the mathematical methods and software used. Several methods are adopted by researchers to explore hidden frequencies in a time series, although it is difficult to decide which one is the "ideal" (Muller and McDonald 2002). Using a combination of frequency analysis and wavelet tools, it is possible to analyze the basic data to explore a band of cycles of one's interest. As an illustration, we show here how cycles centered around a 22-year solar magnetic cycle are highlighted after several filtering steps in a 331-year-long proxy monsoon data (Figure 8.1), which is unevenly spaced, although having near-yearly resolution. For the periodogram, we use codes for unevenly spaced data (Schulz and Stattegger 1997; Schulz and Mudelsee 2002). For the wavelet analysis, we use codes suggested by Torrence and Compo (1998) with Morlet as the basis function. This method is designed for evenly spaced data. We first filter the basic data and then, using an interpolation method, generate evenly spaced data for the subsequent wavelet analysis. Although the raw data is unevenly spaced, the average spacing between two consecutive data points is approximately 1.1 years and, therefore through interpolation, we expect a negligible contribution of spurious signals due to the conversion of the time series from unevenly spaced to evenly spaced data.

Low-pass filters are often used to smooth noisy data because the data are considered to consist of slowly changing variables of interest corrupted by random noise. However, when a datum point is replaced by some kind of local average of the surrounding data points, then the level of noise can be reduced without much biasing to the new values.

We use a Savgol filter (Savitzky–Golay filter; Press et al. 1992) and vary its width and order successively, initially with some trial values, to filter the raw data and see the frequency pattern in the wavelet spectrum. This is a moving window type filter and, instead of averaging, it does least-squares fitting of a polynomial, preserving higher moments and there-fore, manages to track narrower features of a time series. However, it does less smoothening on broad features. The idea behind using this filter is to vary the width and order of the window to suppress the high-frequency signals (analogous to noise) and use that (by subtracting the filtered data from the raw data) further for spectral analysis (Figure 8.3). Earlier, this

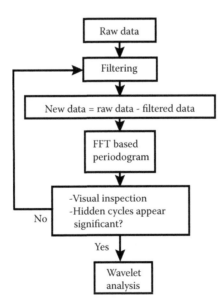

FIGURE 8.3
Flow chart of the method adopted to process the data. Raw data is first filtered with some trial values in the Savgol filter. After each successive filtering, the periodogram is inspected if the dominant low frequency cycles are removed and the high-frequency cycles with low power improve in significance. Wavelet analysis is further taken up or else the data is processed fur-ther with the new values of the Savgol filter.

TABLE 8.1

Savgol Filters of Different Input Parameters Used Here

Filter	Width	Order
F1	15	3
F2	10	4
F3	3	4

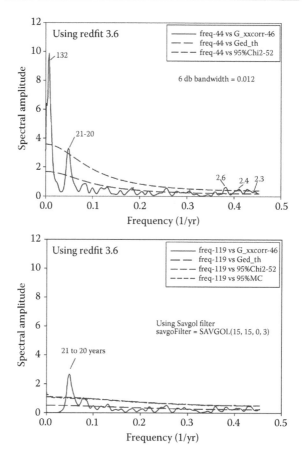

FIGURE 8.4

Effect of filtering on the periodogram is shown here. The raw data are dominated by the low frequency cycles (132 years) (top). After filtering, the low frequency cycles disappear and a cycle of approximately 21 years emerges significantly (bottom).

type of filtering was applied to instrumental rainfall data by Hiremath and Mandi (2004). Several trials were carried out to see the effective width and order of the filter. As the filter does a new least-square fitting each time when the window is moved, the extent to which the high-frequency signal is extracted depends upon the data structure within the window. To carry out repeat analyses, appropriate codes can be developed either in MATLAB® or IDL-based software platforms. Table 8.1 shows the width and the order of the three typical sizes selected for Savgol filters, for which the results are presented here. An example of the effect of filtering on the periodogram is shown in Figure 8.4. Further details are given in Yadava and Ramesh (2007). Figure 8.5 shows the wavelet power spectrum for the raw data, highlighting large contributions of the low period components. When low periods are removed by filter F1, pronounced power seems to be clustered around 22 years during ~1890 CE. By varying the filter parameters, it is possible to further clarify the high-frequency components as shown in Figures 8.6 through 8.8.

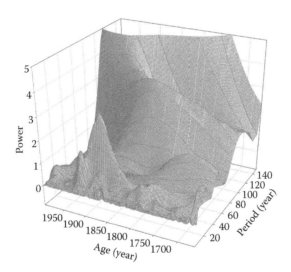

FIGURE 8.5
Wavelet power spectrum of raw data (*z* axis). Period (in years CE) is shown on the *x* axis; *y* axis shows time in years.

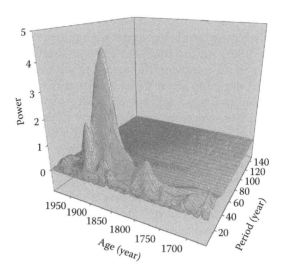

FIGURE 8.6
Wavelet power spectrum after raw data is processed with filter F1, which removes low frequencies and accentuates the band of frequencies that may coincide with that of solar cycle (~22 years). z axis shows power, y axis is the calendar year, and x shows period in years.

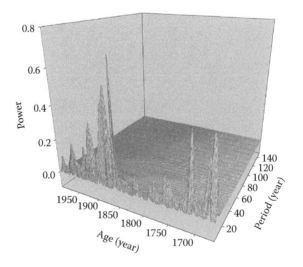

FIGURE 8.7
Filtering with a lower width and higher order filter F2 results in sharper peaks. Axes descriptions are the same as in Figure 8.3.

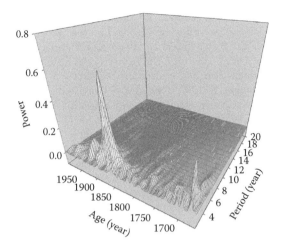

FIGURE 8.8
When the raw data is filtered for low frequencies using filter F3, periods of less than 5 years appear highly accentuated. Effect of order (4 here) can be seen by comparing Figure 8.3 with order 3. Axis descriptions are the same as in Figure 8.3.

Conclusions

We have obtained a strong solar magnetic periodicity of 22 years in the monsoon rainfall data reconstructed from speleothems using wavelets. Multitaper analysis confirms the results and also shows that even carbon isotopes behave in a similar fashion. Furthermore, we explain how, using different filters, we can get hidden signals pertaining to solar modulation of the monsoon.

Acknowledgments

We thank Dr. E. Chandrasekhar, Department of Earth Sciences, IIT Bombay, for his patience and constant encouragement. We thank ISRO-GBP for funding.

References

Blackman, R.B., and Tukey, J.W. 1958. *The Measurement of Power Spectra from the Point of View of Communication Engineering.* Dover Publications, New York, 190 pp.
Hiremath, K., and Mandi, P. 2004. Influence of the solar activity on the Indian Monsoon rainfall. *New Astronomy* 9, 651.

Muller, R.A., and MacDonald, G.J. 2002. *Ice Ages and Astronomical Causes, Data, Spectral Analysis and Mechanisms.* Springer-Verlag, London, 317.

Press, W.H., Teukolsky, S.A., Vetterling, W.T., and Flannery, B.P. 1992. *Numerical Recipes in C,* 2nd ed. Cambridge University Press, Cambridge. p. 650.

Schulz, M., and Mudelsee, M. 2002. *Computers & Geosciences* 28, 421.

Schulz, M., and Stattegger, K. 1997. *Computers & Geosciences* 23, 929.

Thomson, D.J. 1982. Spectrum estimation and harmonic analysis. *Proceedings of the IEEE* 70, 1055–1096.

Torrence, C., and Compo, G.P. 1998. A practical guide to wavelet analysis. *Bulletin of the American Meteorological Society* 79, 61.

Yadava, M.G., and Ramesh, R. 1999. Speleothems—useful proxies for past monsoon rainfall. *Journal of Scientific and Industrial Research* 58, 339–348.

Yadava, M.G., and Ramesh, R. 2007. Significant longer-term periodicities in the proxy record of the Indian monsoon rainfall. *New Astronomy* 12, 544–555.

Yadava, M.G., Ramesh, R., and Pant, G.B. 2004. Past monsoon rainfall variations in peninsular India recorded in a 331 year old speleothem. *The Holocene* 14, 517.

9

Unraveling Nonstationary Behavior in Rainfall Anomaly and Tree-Ring Data: A Wavelet Perspective

Prasanta K. Panigrahi, Yugarsi Ghosh, and Deepayan Bhadra

CONTENTS

Introduction 229
Wavelet Transform: A Brief Introduction.. 230
Discrete Wavelet Transform: An Illustration Using Haar Wavelet 231
Continuous Wavelet Transform and its Uses in Time Series Analysis........ 233
Applications of Wavelets 234
    Analysis of Rainfall Data Using Wavelet Transform................................ 234
        Denoising the Data Using DWT.. 234
        Time Series Plot... 234
        Continuous Wavelet Transform.. 235
        Fourier Analysis.. 237
    Extracting Wavelet Coefficients and Fitting on the Time Series Plot...... 238
    Analysis of Tree-Ring Width Data Using Wavelet Transformation 240
        Detrending Tree-Ring Data 240
        Wavelet Analysis.. 241
Conclusion and Future Work 243
References.. 244

Introduction

We have learned from previous chapters that the advantage of wavelet transform (WT) over Fourier transform (FT) lies in its ability to achieve optimal localization in both time and frequency domains, without violating Heisenberg's uncertainty principle. It is worth mentioning that the scale represents the frequency content, and the finite scale-dependent size of the wavelet endows it with the ability to localize a given frequency modulation of the signals in the time domain (Farge, 1992).

Wavelets are broadly classified as discrete or continuous depending on the span of the basis functions. The discrete wavelet (DW) is composed of a basis set with strictly finite extent, whereas the continuous wavelets (CWs) vanish only at infinity and are usually overcomplete. Although the property of orthogonality and completeness of the DW leads to a completely nonredundant representation of the signals, the CWs such as Morlet, Gaussian, and others have their advantages, as will become clear later in this chapter (Goupillaud et al., 1984; Grossman and Morlet, 1985; Morlet, 1981, 1983; Miao and Moon, 1999). The functions representing CW are oscillatory, with window functions that are scale-dependent. In this respect, they are similar to the "short-time" FT (STFT) or "window FT," which has been used for studying time-varying signals. However, as compared with the windowed FT, the window size of the CW is commensurate with its frequency, which allows us to overcome the drawbacks of STFT.

When time-localization of the spectral components is required, WT is applied, thus providing the time–frequency representation of the signal. For practical applications, particular spectral components occurring in certain time intervals are of interest. It may then help to know the time intervals that these spectral components manifest. WT has gained more prominence because of the shortcomings in STFT's resolution. Wavelets offer variable resolution unlike STFT, higher frequencies are better resolved in time and lower frequencies are better resolved in frequency (Poliker, 1996).

In the following section, we will illustrate the mechanism of the DW through the pictorial representation of wavelets. It is worth mentioning that these Haar wavelet basis sets are the only DWs for which the basic functions can be written in terms of a known function, the Haar function. For all other DW, this is not possible, although one can capture them to any desired accuracy numerically. On a historical note, the Haar wavelet was first invented and employed for a careful analysis and understanding of FT. As the inquisitive ones will find out, FT plays a significant role in the design of the wavelet basis sets.

An amazing aspect of the DW is that although one may not know the basis set, the wavelet transformation and its inverse can be carried out without any approximations whatever. DWs depend only on algebraic operations with its basis functions, satisfying no differential equations. This should be compared with the sine and cosine functions, the basis set of FT that satisfies a second order differential equation originally from Hooke's law.

Wavelet Transform: A Brief Introduction

Mathematical transforms are applied to signals to obtain further information that might not be apparent at first glance (Borgaonkar et al. 1994; Cadet, 1979; Pederson et al., 2001; Singh et al., 2004; Yadava and Ramesh, 2007). Although FTs using the fast Fourier transform (FFT) approach are most commonly used, they can be used only if we are interested in *what* spectral components occur and not *when* they occur. When the time-localization of the spectral components is required, WTs that provide a time–frequency representation of the signal is applied. Wavelets offer variable resolution, higher frequencies are better resolved in time and lower frequencies are better resolved in frequency. WT can be of two types as discussed below.

Discrete Wavelet Transform: An Illustration Using Haar Wavelet

We will illustrate the mechanism of the discrete wavelet transform (DWT) through the Haar wavelet. It is composed of two basic building blocks, the feed scaling functions $\varphi(t)$ (father wavelet) and $\psi(t)$ (mother wavelet). DWT is composed of these two kernels, which satisfy the following admissibility conditions. The father wavelet has a nonzero average, $\int \varphi(t)dt < A$, where A is a constant, whereas $\psi(t)$ is a strictly oscillatory function $\int \psi(t)dt = 0$. φ and ψ are orthogonal, that is, $\int \varphi^*\psi \cdot dt = 0$ and square integrable $\int \varphi^2 \, dt = \int \psi^2 \, dt = 1$. The above are common to all the DWs. The scaling function and translated version of the mother wavelets are explicitly given by $\psi_{j,k}(t) = 2^{\frac{j}{2}}\psi(2^j t - k)$, and are called daughter wavelets with $-\infty \leq k \leq +\infty$ and $j \in \{0, 1, 2, 3,..., \infty\}$. We have taken $A = 1$, a choice resting with the user. Furthermore, these functions are strictly finite and can be written as $\varphi(t) = \theta(t) - \theta(t - 1)$, and $\psi(t) = \theta(t) - 2\theta(2t - 1) + \theta(t - 1)$, where $\theta(t)$ is pictorially represented by $\theta(t) = 1$, for $t > 0$ otherwise zero. The fact that the basis functions have finite size in a time domain spanning the interval $-\infty \leq t \leq +\infty$ necessitates the introduction of a translation operation, which produces an orthogonal set φ_k and ψ_k, through translation in unit steps, the size of the basis functions (Mak, 1995).

The DW decomposition of a function is given by $f(t) = \sum\limits_{k=-\infty}^{\infty} c_k \varphi_k(t) + \sum\limits_{k=-\infty}^{\infty} \sum\limits_{j\geq 0} d_{j,k}\psi_{j,k}(t)$, where c_k are the low-pass coefficients that capture the trend (average behavior or the long wavelength components) and $d_{j,k}$ are the high-pass coefficients that extract the high-frequency components (or fluctuations) of the function. Multiresolution analysis leads to (Mallat, 1989),

$$c_{j,k} = \sum_n h(n - 2k)c_{j+1,n}$$

$$d_{j,k} = \sum_n \bar{h}(n - 2k)c_{j+1,n}$$

where $h(n)$ and $\bar{h}(n)$ are the low-pass and high-pass filter coefficients, respectively. It is worth emphasizing that low-pass and high-pass coefficients at a given scale can be obtained by low-pass coefficients alone at a higher scale.

For illustration, let us consider a data set with four data points. At the lowest scale (i.e., level 1), the low-pass (average) and high-pass (wavelet) coefficients for the data set composed of four data points ($a, b, c,$ and d, respectively) are given by

$$A = \frac{a+b}{\sqrt{2}}, B = \frac{c+d}{\sqrt{2}}, C = \frac{a-b}{\sqrt{2}}, D = \frac{c-d}{\sqrt{2}}$$

Because of Parseval's theorem, the energy/power, that is, $P = a^2 + b^2 + c^2 + d^2$ is conserved.

$$A^2 + B^2 + C^2 + D^2 = \frac{(a+b)^2}{2} + \frac{(a-b)^2}{2} + \frac{(c+d)^2}{2} + \frac{(c-d)^2}{2} = P$$

This explains the origin of the normalizing factor $1/2$ in front of the coefficient. The fact that there are four data points explains the presence of four independent coefficients (two low-pass and two high-pass), from which the data points can be reconstructed by the algebraic operations of addition and subtraction. One observes the absence of coefficients such as $\frac{b+c}{\sqrt{2}}$ and $\frac{b-c}{\sqrt{2}}$, the presence of which would have led to overcounting. A straightforward binning procedure with a window size of two would have yielded the low-pass coefficients.

At the second level, the window size is doubled, which leads to averaging and differentiation at a bigger scale involving four points, to produce the level 2, low-pass and high-pass coefficients:

$$\frac{a+b+c+d}{2} = \frac{A+B}{\sqrt{2}}, \frac{(a+b)-(c+d)}{2} = \frac{A-B}{\sqrt{2}}$$

In general, if the data length is $2N$, one can carry out N levels of decomposition. One notices that the window size is progressively increased in units of two. In the first level decomposition, one obtains low-pass and high-pass coefficients, each of which are exactly half of the data size. In the next level, the level 1 low-pass coefficient is decomposed into level 2 low-pass coefficients and level 2 high-pass coefficients, each half the size of the level 1 low-pass coefficients. This procedure is then carried on to the desired level of decomposition. From the decomposed set of data, the high-pass coefficients can be removed and can be reconstructed by using inverse DWT. The new data set thus obtained will be free from noise. This technique is frequently used for denoising any given set of data before further analysis, and is illustrated later in the chapter.

Continuous Wavelet Transform and its Uses in Time Series Analysis

The Haar wavelet illustrated above is orthogonal in nature. In "time series" analysis of natural data, orthogonal wavelets are not preferred because in orthogonal wavelet analysis the number of convolutions at each scale is proportional to the width of the wavelet basis at that scale. This produces a wavelet spectrum that contains discrete "blocks" of wavelet power and is useful for signal processing because it gives the most compact representation of the signal. Unfortunately, for time series analysis, an aperiodic shift in the time series produces a different wavelet spectrum. Hence, the nonorthogonal transform is useful for time series analysis in which smooth and continuous variations in wavelet amplitude are expected. We therefore use a type of wavelet called the Morlet wavelet, which is continuous and nonorthogonal in nature.

In continuous wavelet transform (CWT), the signal is multiplied with the corresponding wavelet function, and the transform is calculated separately for different segments of the time domain signal. The width of the window is changed as the transform is computed for every scale (spectral component),

$$\text{CWT}_x^\psi(\tau, s) = \psi_x^\psi(\tau, s) = \frac{1}{\sqrt{|s|}} \int x(t) \psi^* \left(\frac{t - \tau}{s} \right) dt$$

The transformed signal is a function of two variables: τ and s, translation and scale parameters, respectively. As noted previously, translation implies that the window is shifted through the signal and scale determines in what detail we view the signal. High scale means a more global view and low scale stands for a more detailed view of the signal. More precisely, wavelet transform is a measure of the frequency similarity between the basis functions (wavelets) and the signal.

If the signal has a major component with the frequency corresponding to a given scale, then the wavelet (the basis function) at the same scale will be close to the signal at the particular location where this frequency component occurs. Therefore, the CWT coefficient computed at this point in the time-scale plane will be a relatively large number.

Applications of Wavelets

Analysis of Rainfall Data Using Wavelet Transform

Climate groups from the United States, the National Oceanic and Atmospheric Administration (NOAA) and the National Climatic Data Center (NCDC), as part of the Global Historical Climate Network, have performed a survey filtering a high-quality rainfall data set for Australia. These data can be downloaded from their web site (http://www.warwickhughes.com/water/links. html). Average annual rainfall data for 130 years (in millimeters) from 1876 to 2005 in the Perth region (West Australia) were collected from the above web site and analyzed using CWT (Morlet wavelet).

Denoising the Data Using DWT

Rain data can be difficult to analyze statistically because it tends to be very "noisy" with large variations between years and, typically, a few years are actually near average (Leuschner and Sirocko, 2003; Neff et al. 2001; Tiwari and Ramesh, 2007). We first normalize the data before our analysis and subject it to a two-level decomposition using DWT with Daubechies 4 wavelet, which separates local linear trends from high-frequency fluctuations (Daubechies, 1992). Then we reconstruct the signal ignoring the high-frequency fluctuations to remove noise (the high-frequency ones). This makes our wavelet scalograms smooth. It also helps in improving the confidence level of various periodicities in our wavelet spectrum. Detailed analyses were done with the help of MATLAB® using the following procedures.

Time Series Plot

Time series data can be loaded from a text file and plotted using MATLAB. It is necessary to normalize the time series before applying wavelet transform on it. The following MATLAB code can be used to load, normalize, and plot a time series data of rainfall anomalies.
Code:

```
z =load('rainfall.txt');%load time series rainfall data from
'.txt' file
n=1876:2005;%130 years on the time axis
%normalizes the rainfall data
m=mean(z);
s=std(z);
i=1:length(z);
z=(z(i)-m)./s;
% plot Time (n) vs. Rainfall (z)
figure(1)
```

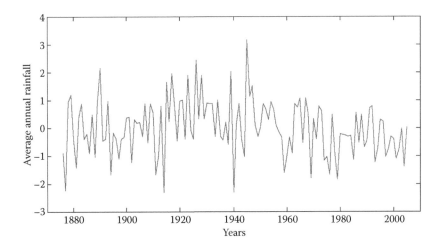

FIGURE 9.1
Normalized plot of rainfall versus years (AD 1876–2005).

```
title('average annual rainfall data(normalized) 1876-2005 A.D);
plot(n,z);
```

The above code generates a normalized time series plot as shown in Figure 9.1.

Continuous Wavelet Transform

Wavelet transform can be applied on any data set using various wavelet windows (e.g., Morlet, Mallat, Gaussian, etc.), which are available in the wavelet toolbox of MATLAB.

The following code helps to plot a three-dimensional representation of CWT coefficients (Figure 9.2), which is obtained using Morlet wavelets. Code:

```
figure();
%3-D plot of wavelet coefficients from scale 1-100
c =cwt(y,1:1:100,'morl','3Dplot');%y is the given time-series
title('3D plot showing wavelet coefficients from scales
1-100');
```

Code:

```
c =cwt(y, 1:100, 'morl');
%cwt is applied on the time series y from scales 1-100
w =wscalogram('image',c);%a scalogram (image) representation
of wavelet coefficients
```

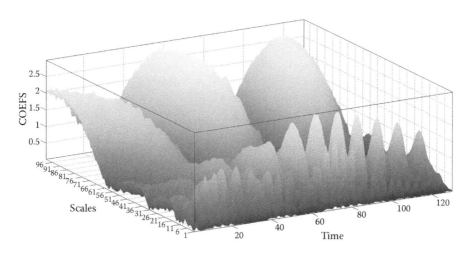

FIGURE 9.2
Three-dimensional plot of the rainfall data scalogram (scale–time versus wavelet coefficients).

As pointed out earlier, it is often necessary to smoothen the time series by removing high-frequency noise before studying it. This can be done by applying DW decomposition up to the desired level of accuracy and reconstructing it after removing the high-frequency (noise) coefficients.

The code below helps to smoothen a time series by removing the high-frequency components.
Code:

```
[c,l] = wavedec(z,2,'db4');
%decomposes to two levels using db4 wavelet
z1 = wrcoef('a',c,l,'db4',2);
%reconstructs only using the low frequency coefficients
```

After smoothening, the time series can be analyzed in a better way because the scalogram of the wavelet coefficients is more prominent. Figure 9.3 compares a normal scalogram (Figure 9.3a) with a smoothened scalogram (Figure 9.3b). Figure 9.3b is plotted after removing the noise components from the time series. It is more prominent and can be perceived and analyzed in a more efficient way.

CWT is applied to the smoothened rainfall data using Morlet wavelet from 0 to 100 scales. Figure 9.2 shows the plot of (scale-time) versus wavelet coefficients, which are obtained by applying CWT on our given set of data. By changing the scales we are able to zoom in and out of our data set and thus capture various periodicities at different resolutions effectively. We observe three strong periodicities of approximately 10, 20, and 30 years' duration.

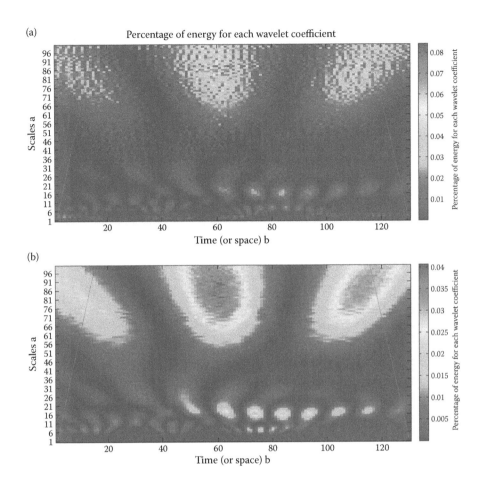

FIGURE 9.3
(See color insert.) (a) Wavelet spectrum of rainfall data without smoothening; (b) wavelet
spectrum of rainfall data after smoothening.

Fourier Analysis

FT gives the frequency domain representation of a time series plot. Using
MATLAB, FT is done efficiently using FFT. It gives the peaks in the frequency domain which are the dominant frequencies present in the time
series.

The following code is used to perform FT on a time series data using FFT
in MATLAB (Oppenheim and Wilsky, 2004; Proakis and Manolakis, 2006).

Code:

```
Fs=1;%sampling rate
T=1/Fs;
L=130;%length of the time series (130 years)
%y is the given time series on which FFT is applied
NFFT = 2^nextpow2 (L);% Next power of 2 from length of y
Y = fft(y,NFFT)/L;
f = Fs/2*linspace(0,1,NFFT/2+1);
figure()
% Plot single-sided amplitude spectrum.
plot ((1./f),2*abs(Y(1:NFFT/2+1)))
title('Single-Sided Amplitude Spectrum of y(t)')
xlabel('time period (years)')
ylabel('|Y(f)|');
xlim([0 100]);
```

The reciprocal of frequency gives the period, which is plotted versus the FFT coefficients. The figure below shows a plot of the dominant periods (1/frequency) present in the time series using FFT.

The presence of the three dominant periodicities of approximately 10, 20, and 30 years in the rainfall data time series is ascertained, as shown in Figure 9.4.

Extracting Wavelet Coefficients and Fitting on the Time Series Plot

From our analysis of rainfall data, we extract three strong periodicities of 10, 20, and 30 years. The confidence levels of the periodicities are tested as stated

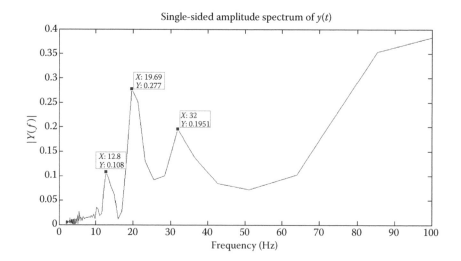

FIGURE 9.4
Plot of $|Y(f)|$ versus time periods ($1/f$) for rainfall data.

by Torrence and Compo (1998). We observe high confidence levels for the three periodicities as shown in Figure 9.5.

Clearly, there are three regions inside the cone of influence (red), corresponding to 10-, 20-, and 78-year periodicities, which show high confidence levels (~90%). Hence, we extract the wavelet coefficients corresponding to the indicated periodicities and analyze them. The 10-year period is of solar origin and is present predominantly in almost all kinds of natural data (Eddy, 1976; Rind, 2002). The 20-year period is extracted and plotted over the data. It fits with good correlation over the data. Hence, we analyze various characteristics after fitting it over the data.

The following code using MATLAB is used to extract wavelet coefficients and plot over the time series.
Code:

```
n=1:130;%length of time series
c=cwt(y, 17, 'morl');
%extracts wavelet coefficients along scale 17 (~20 year period)
plot(n, c);
hold on%pauses and plots one curve over the other
plot(n, y);
```

It is interesting to note that a local minimum of the 20-year period coincides very well with the year 1914 (see Figure 9.6), when Western Australia received the least amount of annual rainfall and faced acute water shortage. It also coincides with the recent scarcity of rainfall in 2001 (Australian Bureau of Meteorology web site for Western Australia (http://www.warwickhughes.com/water/links.html)).

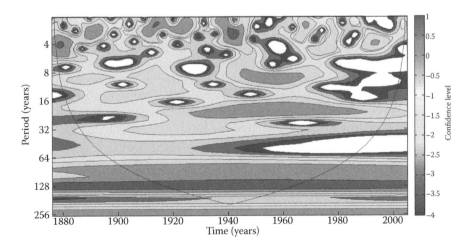

FIGURE 9.5
Wavelet spectrum of rainfall data.

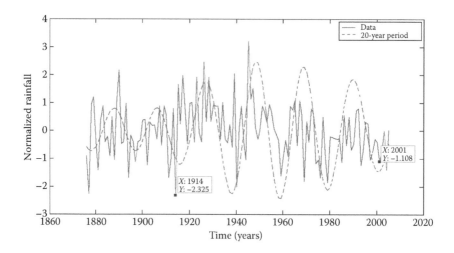

FIGURE 9.6
Plotting 20-year periodicities over the rainfall data.

Analysis of Tree-Ring Width Data Using Wavelet Transformation

Temperature or precipitation data has only been recorded for a short period since the late 1880s or early 1900s. The variations in temperature and precipitation patterns introduce natural variations in the tree-ring width and are thus recorded. Therefore, tree-ring data from AD 1000 are available and they can act as vital proxy sources to characterize periodic modulations for long-term analysis (Castagnoli et al., 1990; Stuiver and Braziunas, 1989). We used data containing raw measurements of tree-ring width of Huon pine (*Lagarostrobos franklinni*) from Buckley Lake region (Western Australia) located at (south −42.33 latitude and east 146 longitude).

Tree-ring data were collected from the web site of Tree-Ring Data–World Data Center for Paleoclimatology (Bruce Bauer, 2012, Personal communication). The International Tree-Ring Data Bank is maintained by the NOAA Paleoclimatology Program and the World Data Center for Paleoclimatology. The data bank includes raw ring width or wood density measurements as well as site chronologies. The data can be downloaded from the web site http://www.ncdc.noaa.gov/paleo/treering.html.

Detrending Tree-Ring Data

The innermost ring widths of a tree are usually also the widest ring widths of that tree. Detrending of a ring width curve is the process of modifying the curve to remove the effects of tree aging from the ring width data. This detrending also includes a scaling of the ring width curve to compensate for the effects of rich or poor soil. A detrended curve contains information about the climate's history, a matter which is the focus of today's research.

The mean of all the tree-ring width series were set to zero for analyzing periodic variations. Because we were interested only in periodic variations, detrending of the time series was done by removal of mean line growth function. Then, they are analyzed using wavelet transform after removing high-frequency fluctuations using Daubechies 4 wavelet.

Wavelet Analysis

We have two sets of data: one ranging from AD 1070 to 1939 (Figure 9.7a) and the other from AD 1590 to 1999 (Figure 9.7b) from the Buckley region in Australia.

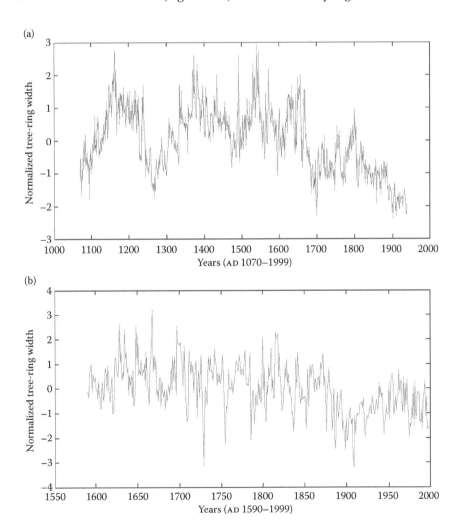

FIGURE 9.7

Variations in tree-ring width of Huon pine (*L. franklinni*) versus year: (a) data set 1 and (b) data set 2.

They are subjected to wavelet transform using Morlet wavelet and the time scale distributions of wavelet coefficients for various resolutions are analyzed carefully.

Periodic modulations of 10, 20, 30, 75, and 170 are found dominant in both data sets, which are also subjected to FT using FFT to confirm the presence of the periodicities.

In our analysis of tree-ring data, we found dominant periodicities of 170, 75, 30, and 20 years (Figure 9.8). We confirm the presence of such periods by analyzing it using FFT (Figure 9.9), as described by the previous MATLAB codes.

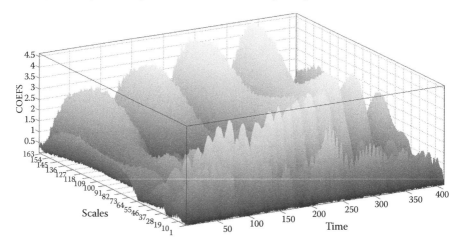

FIGURE 9.8
Three-dimensional plot of the tree-ring widths (data set 1).

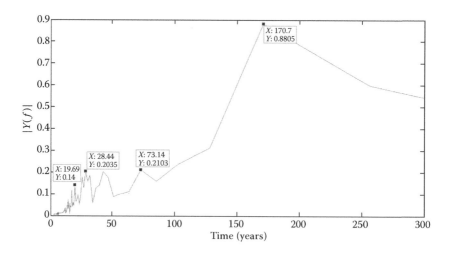

FIGURE 9.9
FT of tree-ring (data set 1).

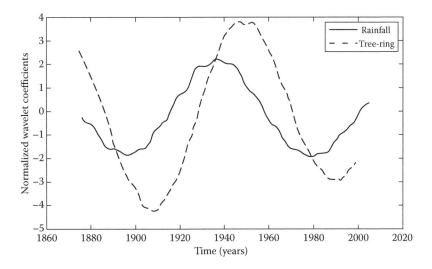

FIGURE 9.10
Comparison of an approximately 75-year period of tree-ring data with a similar period of rainfall data.

The plot in Figure 9.8 clearly indicates the presence of periodic modulations of approximately 20, 30, 73, and 170 years. The peaks indicated according to the FT plot (Figure 9.9) confirm that a periodic modulation near the peak value is present predominantly in the data set. We further ascertain our periodicities by taking different sets of tree-ring data from nearby regions, and similar periodicities are found in all of them. Thus, the various wavelet coefficients corresponding to the periodic modulations are extracted and precisely studied. Now, approximate periodicity of nearly 73 to 78 years is present predominantly in all the three data sets (temperature anomalies, rainfall anomalies, and tree-ring width data). It is clearly seen that the two periodicities (tree-ring width and rainfall; Figure 9.10) are positively correlated with a certain phase shift.

Conclusion and Future Work

Wavelet transform is a useful mathematical tool for extracting periodic modulations because it gives a multiresolution, timescale representation of the signal, unlike FT, which provides only the frequency domain information. Because recorded rainfall data are not available for longer durations, the tree-ring data acts as a significant proxy source for ascertaining periodicities in a longer perspective. A 73- to 78-year period (approximately) is found predominantly in the sets of data. It is found that the rainfall and tree-ring period are significantly correlated (Castagnoli et al., 2002). Hence, wavelet transform can characterize natural phenomena and mathematical analysis of the same can be achieved through MATLAB and can lead to various scientific developments.

References

Borgaonkar, H., Pant, G., and Kumar, K. Dendroclimatic reconstruction of summer precipitation at Srinagar, Kashmir, India, since the late-eighteenth century, *Holocene* 4, 299–306 (1994).

Cadet, D. Meteorology of the Indian Summer Monsoon, *Nature* 279, 761–767 (1979).

Castagnoli, G., Bonino, G., and Taricco, C. Long term solar-terrestrial records from sediments: carbon isotopes in planktonic foraminifera during the last millennium, *Advances in Space Research* 29, 1537–1549 (2002).

Castagnoli, G., Bonino, G., Caprioglio, F., Serio, M., Provenzale, A., and Bhandari, N., The $CaCO_3$ profile in a recent Ionian sea core and the tree ring radiocarbon record over the last two millennia, *Geophysical Research Letters* 17, 1545–1548 (1990).

Daubechies, I. *Ten Lectures on Wavelets, 64 CBMS-NSF Regional Conference Series in Applied Mathematics*. Society for Industrial Mathematics (1992).

Eddy, J.A., The Maunder Minimum, *Science* 192, 1189–1202 (1976).

Farge, M. Wavelet transforms and their applications to turbulence. *Annual Review of Fluid Mechanics* 24(1) 395–458 (1992).

Goupillaud, P., Grossmann, A., and Morlet, J. Cycle-octave and related transforms in seismic signal analysis. *Geoexploration* 23(1), 85–102 (1984).

Grossman, A. and Morlet, J. Decomposition of functions into wavelets of constant shape, and related transforms. In *Mathematics and Physics: Lectures on Recent Results*, edited by L. Streit. World Scientific Publishing, Singapore (1985).

Leuschner, D. and Sirocko, F. Orbital insolation forcing of the Indian. *Palaeogeography, Paleoclimatology, Paleoecology* 197, 83–95 (2003).

Mak, M. Orthogonal wavelet analysis: Inter-annual variability in the sea surface temperature. *Bulletin of the American Meteorological Society* 76, 2179–2186 (1995).

Mallat, S. A theory for multi-resolution signal decomposition: The wavelet representation. *IEEE Transactions on Pattern Analysis and Machine Intelligence* 11(7), 674–693 (1989).

Miao, X.G. and Moon, W.M. Application of wavelet transform in reflection seismic data analysis. *Geosciences Journal*, 3(3), 171–179 (1999).

Morlet, J. Sampling theory and wave propagation. In *Issues in Acoustic Signal/Image Processing and Recognition, NATO ASI Series*, edited by C. Chen. Springer-Verlag, Berlin, pp. 233–261 (1983).

Morlet, J. Sampling theory and wave propagation. In *Proceedings of the 51st Annual Meeting of the Society of Exploration Geophysics (Los Angeles, Society of Exploration Geophysics)*, pp. 1418–1423 (1981).

Neff, U., Burns, S.J., Mangini, A., Mudelsee, M., Fleitmann, D., and Matter, A. Strong coherence between solar variability and the monsoon in Oman between 9 and 6 kyr ago, *Nature* 411, 290–293 (2001).

Oppenheim, A.V., Wilsky, A.S., and Hamid, S. *Signals and Systems.* 2nd edition, Prentice Hall International, NJ (2004).

Pederson, N., Jacoby, G., Arrigo, R.D., Cook, E., Buckley, B., Dugarjav, C., and Mijiddorj, R. Hydrometeorological reconstructions for northeastern Mongolia derived from tree rings: 1651-1995, *Journal of Climate* 14, 872–881 (2001).

Polikar, R. Fundamental Concepts and Overview of Wavelet Theory. 2nd ed. http://users.rowan.edu/~polikar/WAVELETS/WTpart1.html (1996).

Proakis, J.G. and Manolakis, D.G., *Digital Signal Processing.* 4th edition, Prentice Hall, NJ (2006).

Rind, D. The sun's role in climate variations, *Science* 296, 673 (2002).

Singh, J., Yadav, R.R., Dubey, B., and Chaturvedi, R. Millennium-long ring-width chronology of Himalayan cedar from Garhwal Himalaya and its potential in climate change studies, *Current Science* 86, 590 (2004).

Stuiver, M. and Braziunas, T.F. Atmospheric 14C and century-scale solar oscillations, *Nature,* 338, 405–408 (1989).

Tiwari, M. and Ramesh, R. Solar variability in the past and palaeoclimate data pertaining to the southwest monsoon, *Current Science* 93, 477–487 (2007).

Torrence, C. and Compo, G. A practical guide to wavelet analysis. *Bulletin of the American Meteorological Society* 79(1), 61–70, (1998).

Yadava, M. and Ramesh, R. Significant longer-term periodicities in the proxy record of the Indian monsoon rainfall, *New Astronomy* 12, 544–555 (2007).

10

Phase Field Modeling of the Evolution of Solid–Solid and Solid–Liquid Boundaries: Fourier and Wavelet Implementations

M. P. Gururajan, Mira Mitra, S. B. Amol, and E. Chandrasekhar

CONTENTS

Introduction: An Understanding of the Phase Transitions in the Earth 247
Phase Field Modeling 250
    Models 250
    Evolution Equations and their Derivation 251
        Equations for ATG Instabilities 251
        Solving the Equation of Mechanical Equilibrium 254
Spectral Implementation of the Phase Field Models 255
    Nondimensionalization 257
    Results from Spectral Implementation 257
Wavelet Implementation and Benchmarking 259
    Daubechies' Compactly Supported Wavelets 260
    Formulation 261
    Results from Wavelet Implementation (1-D Cahn–Hilliard Equation) 264
Conclusions 268
Appendix A 269
References 270

Introduction: An Understanding of the Phase Transitions in the Earth

The primordial gravitational accretion processes that formed the planet Earth are the basic causes for the increase in temperature, pressure, density, etc., with depth, which gave rise to the modification and separation of the original Earth-forming material into roughly concentric shells. The depth classification of these shells, that is, the crust, the mantle, the outer core, and the inner core is well known. It is important to understand what the boundaries distinguishing these different layers signify. They represent the zones of sudden variation in the physical properties of the media that arise due to solid–solid or solid–liquid phase transitions in the composition of the deep Earth material. The geophysical surface manifestation of such features is discontinuities in seismic wave velocities, electrical conductivity, density, etc., as shown in Figure 10.1. The crust–mantle boundary (known as Mohorovičić discontinuity) and the mantle–core boundary (known as Gutenberg discontinuity) stand as good examples for such solid–solid and solid–liquid transition zones, respectively. Within the upper mantle, the notable discontinuities of geophysical importance are the seismic discontinuities at 410 and 660 km. The correlation of seismic discontinuities and electrical conductivity jumps at these depths and at 800 to 900 km has drawn the attention of many researchers (Bahr et al. 1993; Constable 1993; Olsen 1998; Chandrasekhar 1999, 2000 to cite a few). These zones are usually interpreted as chemical or physical phase transition zones. Laboratory studies further confirm these observations (Omura 1991; Xu et al. 1998; Ohtani and Sakai 2008 and references therein).

Such phase transitions that generally occur due to the prevailing high temperature and pressure conditions at deeper depths represent dynamic and nonequilibrium zones. Furthermore, in many cases, these phase transitions and associated processes can lead to interesting microstructural changes, which in turn control the physical and mechanical properties of the system. A widely used technique for studying such phase transition dynamics and the associated microstructural formation and evolution is the phase field modeling technique, which involves solving nonlinear partial differential equations with relevant boundary conditions. Typically, the phase field models are solved using Fourier spectral techniques. In this chapter, we take the specific problem of a nonhydrostatically stressed mineral in contact with its aqueous solution and show how phase field models can be used to study the same; we use a Fourier spectral method for the implementation of the phase field model. Wavelet-collocation methods can also be used for implementing the phase field models, particularly when the boundary conditions are not periodic; for example, when the system under study is not a representative volume element, in which the periodic boundary condition is not the natural boundary condition. In this chapter, we implement the wavelet-collocation methods for the simplest one-dimensional (1-D) phase field model of the Cahn–Hilliard equation, and compare the results with analytical as well as those obtained from finite difference and Fourier spectral methods.

The organization of the chapter is as follows: We first describe the phase field formulation and its Fourier spectral implementation and present our results from the study of ATG instabilities in a nonhydrostatically stressed mineral in contact with its solution. Then, we present the wavelet formulation for the 1-D Cahn-Hilliard equation and some preliminary results which is followed by our Conclusions.

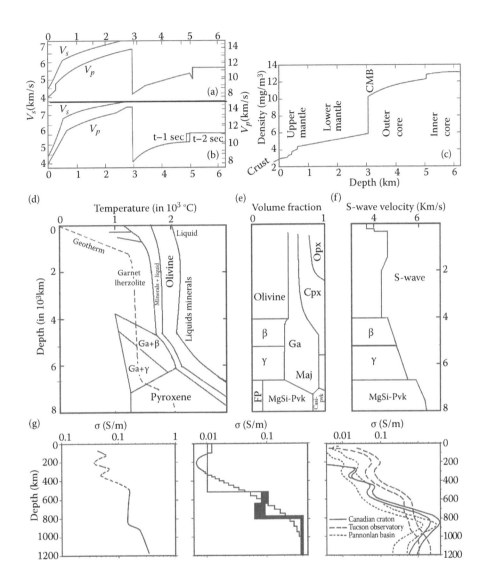

FIGURE 10.1
(a) Velocities of longitudinal (P) and transverse (S) waves in the Earth's interior according to Jefferys (1939) and (b) Gutenberg (1948). (c) Density variations within the Earth as computed from seismic wave velocities. (d) Phase boundaries and geotherm (Ga, garnet, β and γ are forms of spinel). (e) Modal percentages of minerals in upper mantle at temperatures corresponding to the geotherm in (d) Ol, olivine; Opx, orthopyroxene; Cpx, clinopyroxene; Maj, majorite; Perv, perovskite. (After Helffrich, G., and Wood, B., *Geophysical Journal International* 126, F7–F12, 1996.) (f) General seismic velocity profile of S-wave beneath continents. (g) Electrical structure of the mantle for the Indian region (left panel), the European region (middle panel). (After Olsen, N., *Geophysical Journal International* 133, 298–308, 1998.) and other global regions (right panel). The conductivity increase at 850 km depth range, depicting the presence of a *midmantle conductor* is considered to be a global phenomenon. (From Chandrasekhar, E., *Earth, Planets and Space* 52, 587–594, 2000.)

Phase Field Modeling

Phase field models, which are also known as diffuse interface models, have been used quite successfully in the past few decades to study phase transformations and the microstructures that result from the phase transformations [see Chen (2002) and Boettinger et al. (2002) for an overview]. The most important advantage of phase field models is that an explicit tracking of interfaces is not required in these models, unlike the sharp interface models, in which the interfaces are tracked. In the latter, in addition to tracking of interfaces, the relevant boundary conditions need to be implemented at the interfaces. This helps in the automatic incorporation of interface-related effects (such as the Gibbs–Thomson effect) and topological singularities in which either new interfaces form or older ones disappear, making it easy to model and study such features numerically.

Models

In phase field models, the microstructure that is under study is described by field variables known as order parameters—for example, a microstructure with two different phases that differ in their composition is described by the composition field, and a microstructure of a solid in contact with its melt is described by an order parameter (known as phase field), which distinguishes a solid from liquid by assigning two different values in the two phases. Note that the composition field is a conserved order parameter (i.e., the integral of the order parameter over the entire system remains a constant throughout the simulations), whereas the phase field is a nonconserved order parameter (i.e., the integral of the order parameter over the entire system keeps changing with time during the simulation). Once the order parameter is determined, the thermodynamics of the system is described (i.e., the free energy in the case of isothermal systems and entropy in the case of nonisothermal systems) in terms of these order parameters and their gradients. One important way in which phase field models differ from their sharp interface counterparts (the classic diffusion equation, Fourier law of heat conduction and so on) is that the thermodynamics are described in terms of functionals of order parameters instead of functions of order parameters. Finally, using variational principles (for the extremization of free energy/entropy), we derive the equations for the evolution of the microstructures. Thus, the phase field models (by definition) are thermodynamically consistent. Furthermore, we also introduce conservation laws if any (mass or energy conservation), while writing the evolution equations. In addition, there could be additional equations which are to be solved; in the case of elastically stressed solids, for example, the equation of mechanical equilibrium is to be solved to incorporate the stresses and strains into the free energy. In the next section, we show an example of the way in which phase field evolution equations are set up.

Evolution Equations and their Derivation

Asaro–Tiller–Grinfeld instabilities (ATG instabilities) are instabilities that occur in interfaces between nonhydrostatically stressed solids and compliant phases. They were predicted (independently) by Asaro and Tiller in 1972 in the context of stress corrosion cracking using stability analysis (Asaro and Tiller 1972) and by Grinfeld in 1986 (Grinfeld 1986) based on variational arguments. ATG instabilities have been observed at many different length scales in many different systems: for example, they have been observed in He-IV solid–liquid interfaces (Torii and Balibar 1992), in SiGe thin films on Si (Pidduck et al. 1992), and in the case of polymer single-crystal films (Berrehar et al. 1992). ATG instabilities have also been observed experimentally in the case of stressed minerals in contact with their aqueous solutions (see Morel and den Brok 2001; den Brok et al. 2002; Koehn et al. 2003) and thus they might control the structure of grain boundaries of wet minerals and hence the mechanical behavior of the Earth's crust (den Brok et al. 2002). Koehn et al. (2003) have tried to model the ATG instabilities in such systems that undergo stress corrosion cracking using a combination of discrete element and continuum models. In this section, we show how phase field models can be set up and how evolution equations can be derived to study the early stages of ATG instabilities in systems that undergo such stress corrosion cracking.

Equations for ATG Instabilities

The formulation and spectral implementation of a phase field model for studying microstructures and their evolution in elastically stressed (and elastically anisotropic and inhomogeneous) systems is described in detail in Gururajan (2006), Gururajan and Abinandanan (2007), and Chirranjeevi et al. (2009). However, for the sake of completion, we describe the formulation in this subsection and the details of the nondimensionalization and numerical implementation in the next section.

Let us consider a stressed solid in contact with its aqueous solution as shown in Figure 10.2. As stated earlier, the first step in setting up of a phase field model is the identification of the order parameters. In this case, it is clear

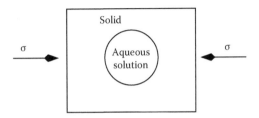

FIGURE 10.2
Schematic of a solid in contact with its solution under compressive stresses in the horizontal direction.

that we need two order parameters. One is the phase field order parameter, which indicates whether the given grid point at any time is occupied by a solid or its aqueous solution. The second is the composition order parameter, which indicates the amount of solute in the solution. However, for the sake of simplicity, we use only the composition order parameter in this chapter and use the composition values themselves as proxies for the phase. The composition parameter used by us in this formulation is scaled to be in the range zero to unity (in the absence of interface and elastic stress–related Gibbs–Thomson effects). The formulation and implementation of a phase field model with both composition and order parameters to study ATG instabilities is in progress and will be reported elsewhere.

Once the order parameter is identified, the next step is to write the free energy of the system in terms of the order parameters and their gradients. In the present problem, the free energy functional consists of two parts, namely, the chemical free energy and the elastic energy. The chemical free energy itself consists of two parts: the bulk free energy density (which is a double well potential), which accounts for the fact that two phases, namely, the aqueous solution and the solid, are in contact with each other at equilibrium and the gradient term, which accounts for the interface. The elastic energy is given by the usual expression,

$$\frac{1}{2}\int_V \sigma{:}\varepsilon \ dV \qquad (10.1)$$

where σ is the local stress, ε is the local strain, and V is the domain volume. Thus, the free energy density becomes

$$F = F^{ch} + F^{el} = N_V \int \left(Ac^2(1-c)^2 + \kappa(\nabla c)^2 \right) dV + \frac{1}{2}\int \sigma{:}\varepsilon \, dV \qquad (10.2)$$

where N_V is a Avogadro number, A is a constant related to the free energy barrier between the two phases, c is the (scaled) local composition, and κ is the gradient energy coefficient that determines the interfacial energy and thickness (along with A).

Note that the elastic stresses are calculated using the elastic moduli, which are different in the two phases. Thus, the elastic energy is an implicit function of the order parameter.

Let us assume the following:

(1) The time scales associated with compositional changes are far smaller than those associated with elastic relaxation and hence these two can be decoupled; in other words, for a given composition profile, the equation of mechanical equilibrium ($\nabla{\cdot}\sigma = 0$) is solved and the stress and strain fields can be obtained, which can then be used to drive the evolution of the composition field.

(2) The solid obeys Hooke's law of elasticity. We also assume the aqueous solution to behave like a Hookean solid albeit with a shear modulus of nearly zero. As we show later, the solution behaves like a fluid in the sense that the stress field inside the fluid for our choice of elastic moduli is almost hydrostatic.

(3) The dependence of the elastic moduli on composition is given by the expression as follows:

$$C_{ijkl}(c) = \frac{1}{2}(C^{\alpha}_{ijkl} + C^{\beta}_{ijkl}) + [c^3(10 - 15c + 6c^2) - 0.5](C^{\alpha}_{ijkl} - C^{\beta}_{ijkl}) \qquad (10.3)$$

where α and β represent the two phases and C_{ijkl} is the elastic moduli. Note that for two different values of composition, 0 and 1, which correspond to α and β phases, respectively, the corresponding moduli is obtained. For any intermediate values of composition, the moduli is interpolated using a function that is continuous and smooth.

(4). The rate of change of the order parameter is proportional to the chemical potential (this is Fick's first law). The chemical potential is given by the variational derivative of the free energy with respect to the order parameter (see, for example, Hilliard 1970). This assumption along with the law of conservation of mass, directly leads to the following evolution equation:

$$\frac{\partial c}{\partial t} = \nabla M \nabla \mu \qquad (10.4)$$

where M is the mobility and μ is the chemical potential related to the free energy as follows:

$$\mu = -\frac{\delta\left[\dfrac{F}{N_V}\right]}{\delta c} \qquad (10.5)$$

where $(\delta/\delta c)$ is the variational derivative with respect to c.

Thus, from Equations 10.2 and 10.5, we see that the chemical part of the free energy gives the following contribution to the chemical potential:

$$\mu^{ch} = h - 2\kappa\nabla^2 c \qquad (10.6)$$

where $h = 2Ac(1 - c)(1 - 2c)$

Solving the Equation of Mechanical Equilibrium

To calculate the contribution of the elastic part of the free energy to the chemical potential, let us consider the elastic energy expression in Equation 10.1. Using Hooke's law, and assuming that the strain is compatible with the displacement field **u**, one can obtain

$$F^{el} = \frac{1}{2}\int \sigma{:}\varepsilon \, dV = \frac{1}{2}\int C_{ijkl}(c)\left[\frac{\partial u_i}{\partial r_j}\frac{\partial u_k}{\partial r_l}\right] \tag{10.7}$$

where **r** is the position vector. Thus, to calculate the elastic energy, one has to solve the equation of mechanical equilibrium, which in terms of displacement field becomes

$$\frac{\partial^2 (C_{ijkl}u_k)}{\partial r_j \partial r_l} = 0 \tag{10.8}$$

Note that the equation of mechanical equilibrium is to be solved assuming different elastic constants for the two different phases; that is, it is a homogenization problem. Furthermore, also note that our system is periodic; that is, the composition is periodic and hence the elastic moduli are periodic. Furthermore, the entire macroscopic system could be under an homogeneous applied stress of σ^A; that is, despite the local heterogeneities, the system behaves as if it is a homogeneous block. This condition leads to antiperiodicity of applied traction (because the normals to the boundaries are of the opposite sign).

The periodicity of the composition field and elastic moduli imply that in the solution to the equation of mechanical equilibrium, the elastic stresses should be periodic. However, we calculate the displacements while solving the equation of mechanical equilibrium. Hence, in terms of displacements, we are looking for solutions to the displacement field that are strain periodic:

$$u = E{\cdot}r + u^* \tag{10.9}$$

where u^* is the periodic displacement field and E is a constant homogeneous strain tensor (and can be shown to be the mean strain tensor for the cell). Finally, the mean stress over the cell is equal to the applied stress σ^A:

$$\sigma^A = \frac{1}{V}\int \sigma \, dV = \frac{1}{V}\int C_{ijkl}(E_{kl} + \varepsilon_{kl}^*) \tag{10.10}$$

where ε^* is the periodic strain defined as follows:

$$\varepsilon_{ij} = \frac{1}{2}\left(\frac{\partial u_i}{\partial r_j} + \frac{\partial u_j}{\partial r_i}\right) \tag{10.11}$$

Thus, solving the equation of mechanical equilibrium reduces to the following problem:
Solve

$$\frac{\partial}{\partial r_j}\left[C_{ijkl}(E_{kl}+\varepsilon_{kl}^*)\right]=0 \tag{10.12}$$

subject to the constraint:

$$E_{ij}=S_{ijkl}\left(\sigma_{kl}^A-\left\langle\sigma_{kl}^*\right\rangle\right) \tag{10.13}$$

where

$$\left\langle\sigma_{ij}^{el}\right\rangle=\frac{1}{V}\int_\Omega\sigma_{ij}^{el}\,d\Omega\,;\text{ that is, }\sigma_{kl}^A=\frac{1}{V}\int_\Omega C_{ijkl}(E_{kl}+\varepsilon_k^*)d\Omega$$

and

$$S_{ijkl}=(<C_{ijkl}>)^{-1} \tag{10.14}$$

In other words, we achieve stress control through strain control while solving the equation of mechanical equilibrium. As a result of all the above calculations, one obtains the chemical potential due to the elastic energy as

$$N_V\mu^{el}=\frac{1}{2}\theta(c)\left(C_{ijkl}^\alpha-C_{ijkl}^\beta\right)\left(E_{ij}+\varepsilon_{ij}^*\right)\left(E_{kl}+\varepsilon_{kl}^*\right) \tag{10.15}$$

where

$$\theta(c)=\frac{\partial}{\partial c}[c^3(10-15c+6c^2)] \tag{10.16}$$

Spectral Implementation of the Phase Field Models

It is possible to solve the equation of mechanical equilibrium using an iterative procedure based on Fourier transforms [see Khachaturyan et al. (1995), Michel et al. (1999), and Hu and Chen (2001)]. The phase field equation for composition can also be solved using the Fourier spectral technique. In this subsection, as noted earlier, for the sake of completion, we describe the Fourier transform–based spectral technique for solving these equations.

Consider the equation of mechanical equilibrium as shown in Equation 10.12 along with Equation 10.11:

$$\frac{\partial}{\partial r_j}\left(C_{ijkl}(c)\left[E_{kl} + \frac{\partial u_k}{\partial r_i}\right]\right) \qquad (10.17)$$

Consider rewriting Equation 10.3 as follows:

$$C_{ijkl}(c) = C_{ijkl}^{eff} + \eta(c)\,\Delta C_{ijkl} \qquad (10.18)$$

where

$$C_{ijkl}^{eff} = \frac{1}{2}\left(C_{ijkl}^{\alpha} + C_{ijkl}^{\beta}\right) \qquad (10.19)$$

$$\eta(c) = c^3(10 - 15c + 6c^2) - 0.5 \qquad (10.20)$$

and

$$\Delta C_{ijkl} = C_{ijkl}^{\alpha} - C_{ijkl}^{\beta} \qquad (10.21)$$

Hence, Equation 10.17 becomes

$$\frac{\partial}{\partial r_j}\left(\left[C_{ijkl}^{eff} + \eta(c)\,\Delta C_{ijkl}\right]\left[E_{kl} + \frac{\partial u_k}{\partial r_i}\right]\right) \qquad (10.22)$$

In other words,

$$\left[C_{ijkl}^{eff}\frac{\partial^2}{\partial r_j\partial r_i} + \Delta C_{ijkl}\left(\frac{\partial}{\partial r_j}\eta(c)\frac{\partial}{\partial r_i}\right)\right] = -\Delta C_{ijkl}E_{kl}\frac{\partial\eta(c)}{\partial r_j} \qquad (10.23)$$

Equation 10.23 can be solved numerically, using a Fourier transform–based iterative technique. As the zeroth order approximation, assume $\Delta C = 0$. Then, Equation 10.23 becomes

$$C_{ijkl}^{eff}\frac{\partial^2 u_k}{\partial r_j\partial r_l} = 0 \qquad (10.24)$$

The above equation can be solved using Fourier transforms. For the next iteration, this solution is used as the starting point as well as to calculate the homogeneous strain (using Equation 10.13) so that Equation 10.23 can be solved including ΔC terms—again using Fourier transforms.

We use the given composition profile to solve the equation of mechanical equilibrium; we then calculate both the chemical and elastic part of the chemical potentials; using these chemical potentials in Equation 10.4, and using a semi-implicit Fourier spectral method, we obtain the composition profile at the next time step. The relevant chemical potentials and the solution for the new time are obtained. We keep marching in time in this manner to obtain the microstructural evolution.

Nondimensionalization

In all our simulations reported here, we have used nondimensional parameters. We carry out the nondimensionalization by using a length scale of $\left(\dfrac{\kappa}{A}\right)^{\frac{1}{2}}$, an energy scale of A, and a time scale of $\dfrac{\kappa[C^\alpha - C^\beta]^2}{AM}$ where A is the constant from bulk free energy density, κ is the gradient energy coefficient, M is the mobility, and C^α and C^β are the compositions of the α and β phases, respectively. These choices result in a value of unity for A, κ, and M. Furthermore, with these nondimensionalizations, in a system with an interfacial energy of 0.1 J m^{-2}, a dimensional shear modulus of 400 G Pa will be converted to a nondimensional shear modulus of 300.

Results from Spectral Implementation

In the simulation results presented here, we assume that both the solid and the liquid are isotropic. We further assume the elastic constants (nondimensional) of $c_{11} = 80$, $c_{12} = 40$, and $c_{44} = 20$ for the solid and $c_{11} = 12.01$, $c_{12} = 12.0$, and $c_{44} = 0.0005$ for the liquid. The (nondimensional) parameters used in the simulation are listed in Table 10.1.

As noted earlier, we have used the composition itself as a proxy for distinguishing the solid phase from the liquid phase. We achieved this through the appropriate choice of the elastic moduli for the liquid phase: that is, the chosen elastic moduli are isotropic and the shear modulus is negligible. In Figure 10.3, we show the σ_{11}, σ_{22}, and σ_{12} along the diagonal line across the simulation domain at time 100; for the sake of clarity, we also show the composition profile along the same diagonal line at the right-hand side of the figure. From these plots, it is clear that inside the liquid ($c = 1$), only hydrostatic stress is supported (that is, as expected, $\sigma_{11} = \sigma_{22}$, and $\sigma_{12} = 0$), whereas the solid ($c = 0$) supports both nonhydrostatic and shear stresses.

In Figure 10.4, we show the microstructures at three different times in a simulation done on a square domain of size 1024×1024. The circular hole at

TABLE 10.1

The Model, Simulation, and Elastic Parameters Used in our Simulations

Parameter Type	Parameter	Value
Cahn–Hilliard Model	M	1
Cahn–Hilliard Model	κ	1
Cahn–Hilliard Model	A	1
Simulation	Δx	0.5
Simulation	Δy	0.5
Simulation	Δt	0.05
Simulation	Allowed error in displacements	0.00001
Simulation	Number of nodes in x direction	1024
Simulation	Number of nodes in y direction	1024
Elastic	Applied stress along x axis	−3.0

the center is of diameter 300. The assembly is compressed along the x axis (with a compressive stress of −3). The microstructures shown correspond to times of 100, 5000, and 19,000; these three times, respectively, correspond to (a) initial microstructure, (b) the onset of instability, and (c) the late stage evolution of the microstructure. As is clear from the figures, the solid–liquid boundary is unstable and leads to branching, which marks the beginning of the ATG instability. What is interesting is that even though the assembly is compressed only in the x direction, the branching happens both in the x and y directions (even though, at the intermediate stage, the branching is seen only in the y direction).

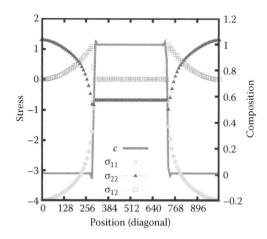

FIGURE 10.3
The stresses along the diagonal of the simulation domain at time 100. The composition is shown on the right-hand side. Inside the liquid (denoted by $c = 1$), the shear stresses are zero and the stress state is hydrostatic. On the other hand, in the solid (denoted by $c = 0$), the two principal stresses are not equal and the shear stress is also not uniformly zero.

FIGURE 10.4
The microstructures of the stressed mineral (compressive stress along the x direction) in contact with its solution at three different times, namely, 100, 5000, and 19,000, showing the onset of ATG instabilities. The microstructure at $t = 100$ is the initial one; by $t = 5000$, the instabilities have set in, and the microstructure at $t = 19,000$ shows the late stage evolution of the instabilities.

Our calculations are carried out in the elastic regime and the microstructural evolution is assumed to be only through diffusion; hence, we do not see the formation of cracks that is observed in the experiments and certain FEM simulations of den Brok et al. (2002) and Koehn et al. (2003).

Wavelet Implementation and Benchmarking

In the results described in the previous section, we have implicitly assumed periodic boundary conditions by virtue of using the Fourier spectral technique. However, in many cases, where one is interested in the elastic stress effects in a system rather than on a representative volume element of a microstructure, it is important to use nonperiodic boundary conditions. Wavelet-collocation methods can be used to study such systems (Gopalakrishnan and Mira 2010). Furthermore, they give the advantage of both spectral and collocation techniques (namely, accuracy). Wavelet implementations of phase field methods are also reported to be more efficient (a) if certain approximations (namely, coefficient filtering; that is, throwing away the coefficients in the wavelet transform that are small, which in effect means that the calculations are only carried out in regions where the gradients are appreciable) can be made (Wang and Pan 2004) and (b) if wavelets can be used to carry out multiresolution calculations (Cogswell 2010). In this section, we describe the wavelet-collocation formulation for the simplified case of a 1-D system without elasticity. The solution obtained for such a case is compared with the analytical results as well those obtained using finite difference and Fourier spectral methods. Our preliminary results indicate that wavelet-collocation implementations of phase field models can be promising.

Daubechies' Compactly Supported Wavelets

A brief review of orthogonal compactly supported basis of Daubechies wavelets (Daubechies 1992) is provided in this subsection. Wavelets $\psi_{j,k}(t)$ form a compactly supported orthonormal basis for $L^2(R)$. The wavelets and the associated scaling functions $\varphi_{j,k}(t)$ are obtained by translation and dilation of single functions $\psi(t)$ and $\varphi(t)$, respectively.

$$\psi_{j,k}(t) = 2^{\frac{j}{2}}\, \psi(2^j t - k)\ j, k \in Z \tag{10.25}$$

$$\varphi_{j,k}(t) = 2^{\frac{j}{2}} \varphi(2^j t - k)\ j, k \in Z \tag{10.26}$$

The scaling functions (t) are derived from the dilation or scaling equation using the recursive relation given by

$$\varphi(t) = \sum_k a_k \varphi(2t - k) \tag{10.27}$$

and similarly, the wavelet function $\psi(t)$ is obtained as

$$\psi(t) = \sum_k (-1)^k a_{1-k} \varphi(2t - k) \tag{10.28}$$

where a_k are the filter coefficients and they are fixed for a specific wavelet or a scaling function basis. For compactly supported wavelets, only a finite number of a_k are nonzero. The filter coefficients a_k are derived by imposing certain constraints on the scaling functions, which are as follows: (1) the area under scaling function is normalized to one, (2) the scaling function (t) and its translates are orthonormal, and (3) the wavelet function $\psi(t)$ has M vanishing moments. The number of vanishing moments M denotes the order N of the Daubechies wavelet, where $N = 2M$ (see Chapter 1 for more details on vanishing moments).

Let $P_j(f)(t)$ be the approximation of a function $f(t)$ in $L^2(R)$ using $\varphi_{j,k}(t)$ as basis, at a certain level (resolution) j, then

$$P_j(f)(t) = \sum c_{j,k} \varphi_{j,k}(t)\ k \in Z \tag{10.29}$$

where $c_{j,k}$ are the approximation coefficients. In this work, a similar approximation of the variables is done using Daubechies scaling function approximation.

Formulation

Here, the 1-D nonlinear diffusion (Cahn–Hilliard) equation is solved using Daubechies scaling function as the approximation bases. Unlike the corresponding Fourier transform–based solution, wavelet transform allows the solution of the problem for nonperiodic boundary conditions. This is primarily because of the compact support of the Daubechies wavelet bases, which helps in finite domain analysis. The 1-D Cahn–Hilliard equations is given as,

$$\frac{\partial c}{\partial t} = \frac{\partial}{\partial x}\left(M\frac{\partial \mu}{\partial x}\right) \tag{10.30}$$

where M is the mobility, $c(r,t)$ is the composition at a time instant t and at location r, and μ is the chemical potential considered as,

$$\mu = h - 2\kappa \frac{\partial^2 c}{\partial x^2} \tag{10.31}$$

where κ is the gradient energy coefficient and

$$h = 2Ac(1 - c)(1 - 2c) \tag{10.32}$$

where A is a positive constant indicating the energy barrier between the two equilibrium phases.

The mobility M and the gradient energy κ are taken as constants assuming the interfacial energies and diffusivities to be isotropic.

Using Equation 10.31 and assuming M as constant, the Cahn–Hilliard equation can be rewritten as,

$$\frac{\partial c}{\partial t} = M\frac{\partial^2}{\partial x^2}\left(h - 2\kappa \frac{\partial^2 c}{\partial x^2}\right) \tag{10.33}$$

For a given initial composition, $c(r,0)$, Equation 10.33 is solved to obtain the composition at any time instant t.

The first step in the wavelet-based solution scheme is the spatial approximation of the composition $c(r,t)$ using Daubechies scaling functions. Let $c(r,t)$ be spatially discretized at m points in the domain $[0, R]$. Let $\xi = 0, 1, 2,..., m - 1$ be the spatial sampling points and

$$r = \Delta r \xi \tag{10.34}$$

where Δr is the sampling interval. The function $c(r,t)$ can be approximated by scaling function $\varphi(\xi)$ at an arbitrary scale as,

$$c(r,t) = c(\xi,t) = \sum_k \hat{c}_k(t)\varphi(\xi - k) \, k \in Z \tag{10.35}$$

where $\hat{c}_k$ are the approximation coefficients at a certain time instant t. Similarly, $h(r,t)$ can be approximated as

$$h(r,t) = h(\xi,t) = \sum_k \hat{h}_k \varphi(\xi - k) \, k \in Z \tag{10.36}$$

Substituting Equations 10.35 and 10.36 in Equation 10.33, we get

$$\sum_k \frac{\partial \hat{c}_k}{\partial t} \varphi(\xi - k) = M \sum_k \frac{1}{\Delta r^2} \hat{h}_k \varphi''(\xi - k) - \frac{2}{\Delta r^4} \hat{c}_k \varphi''''(\xi - k) \tag{10.37}$$

Taking the inner product on both sides of Equation 10.37 with $\varphi(\xi - j)$, where $j = 0, 1, 2, \ldots, m - 1$, we get,

$$\sum_k \frac{\partial \hat{c}_k}{\partial t} \int \varphi(\xi - k)\varphi(\xi - j)d\xi =$$

$$M \sum_k \left(\frac{1}{\Delta r^2} \hat{h}_k \int \varphi''(\xi - k)\varphi(\xi - j)d\xi - \frac{2\kappa}{\Delta r^4} \hat{c}_k \int \varphi''''(\xi - k)\varphi(\xi - j)d\xi \right) \tag{10.38}$$

The translations of scaling functions are orthogonal, that is

$$\int \varphi(\xi - k)\varphi(\xi - j) = 0 \text{ for } j \neq k \tag{10.39}$$

Using Equation 10.39, Equation 10.38 can be rewritten as

$$\frac{\partial c_j}{\partial t} = M \sum_{k=j-N+2}^{j+N-2} \left(\frac{1}{\Delta r^2} \hat{h}_k \Omega_{j-k}^2 - \frac{2\kappa}{\Delta r^4} \hat{c}_k \Omega_{j-k}^4 \right) \quad j = 0, 1, 2, \ldots m - 1 \tag{10.40}$$

where N is the order of the Daubechies scaling function as discussed earlier. Ω_{j-k}^2 and Ω_{j-k}^4 are second-order and fourth-order connection coefficients and are defined as,

$$\Omega^2_{j-k} = \int \varphi''(\xi - k)\varphi(\xi - j)d\xi \qquad (10.41)$$

$$\Omega^4_{j-k} = \int \varphi''''(\xi - k)\varphi(\xi - j)d\xi \qquad (10.42)$$

For compactly supported wavelets, the connection coefficients, Ω^2_{j-k} and Ω^4_{j-k} are nonzero only in the interval $k = j - N + 2$ to $k = j + N - 2$. The details for evaluation of connection coefficients for different order of derivatives are given by Beylkin (1992).

It can be observed from Equation 10.40 that certain approximation coefficients $\hat{c}_j$ and $\hat{h}_j$ in the vicinity of the boundaries ($j = 0$ and $j = m - 1$) lie outside the finite domain defined by $[0, R]$. For finite domain analysis, these coefficients are treated using a wavelet extrapolation technique (Williams and Amaratunga 1997), following which Equation 10.40 can be written in matrix form as,

$$\frac{\partial \tilde{c}_j}{\partial t} = M\left(\Gamma_2\{\tilde{h}_j\} - 2\kappa\Gamma_4\{\tilde{c}_j\}\right) \qquad (10.43)$$

where the matrices, Γ_2 and Γ_4 are the matrices containing the second-order and fourth-order connection coefficients, Ω^2_{j-k} and Ω^4_{j-k}, respectively. Again, for zero initial condition, the higher order connection coefficient matrices can be written as,

$$\Gamma_2 = \Gamma_1^2 \text{ and } \Gamma_2 = \Gamma_1^4 \qquad (10.44)$$

where Γ_1 is the connection coefficient matrix for the first-order derivative. Using, Equation10.39 in Equation 10.43, we get

$$\frac{\partial \tilde{c}_j}{\partial t} = M\left(\Gamma_1^2\{\tilde{h}_j\} - 2\kappa\Gamma_1^4\{\tilde{c}_j\}\right) \qquad (10.45)$$

It should be noted that the equations given by Equation 10.45 are coupled, unlike the corresponding Fourier transform–based solutions. However, the system of equation can be decoupled by diagonalizing the connection coefficient matrices as

$$\Gamma_1 = \Phi\Pi\Phi^{-1} \qquad (10.46)$$

where Φ is the eigenvector matrix of Γ_1 and Π is the diagonal eigenvalue matrix containing the corresponding eigenvalues λ_j. The decoupled equations can be written as

$$\frac{\partial \tilde{c}_j}{\partial t} = \lambda_j^2 M \tilde{h}_j - 2\kappa\lambda_j^4 \tilde{c}_j \qquad (10.47)$$

where $\tilde{c}_j = \Phi^{-1}\hat{c}_j$ and $h_j = \Phi^{-1}\hat{h}_j$.

The temporal discretization of Equation 10.47 is then obtained as

$$\frac{\tilde{c}_j(t+\Delta t) - \tilde{c}_j(t)}{\Delta t} = \lambda_j^2 M \tilde{h}_j(t) - 2\lambda_j^4 \tilde{c}_j(t+\Delta t)$$

$$\tilde{c}_j(t+\Delta t) = \frac{\tilde{c}_j(t) - \lambda_j^2 \tilde{h}_j \Delta t}{1 + 2\Delta t \lambda_j^4} \qquad (10.48)$$

where Δt is the time step. $c(r,t)$ can be obtained from $c_j(t)$ through inverse transform.

Results from Wavelet Implementation (1-D Cahn–Hilliard Equation)

Consider the 1-D Cahn–Hilliard equation (i.e., Equation 10.9) with the boundary conditions that at the left end ($x = 0$), $c = 0$ and at the right end ($x = L$), $c = 1$ (see the step input profile in Figure 10.6) and an initial profile in which at around $x = L/2$, the composition is ramped from zero to unity over 10 or so node points. For the given boundary and initial conditions, there exists an analytical solution for the problem, which is given by $\frac{1}{2}\left[1 + \tanh\left(\frac{x}{2\sqrt{\frac{\kappa}{A}}}\right)\right]$.

In Figure 10.5, we compare the numerical solutions obtained using the wavelet, finite difference, and Fourier transform implementations with the corresponding analytical solution. Note that in the Fourier implementation, we need to use double the system size and two interfaces (by mirroring the profile about L) to solve the problem consistent with the implicit periodic boundary conditions. From the figure, it is clear that all three methods give correct solutions. In Table 10.2, we list the L^2 norm errors (obtained by taking the difference between the analytical solution and the numerical solution, squaring the difference, adding the errors for the interface nodes, dividing the sum by the total number of node points, and taking the square root) for the three different methods for different simulation parameters. From the errors, it is clear that the wavelet solutions are the most accurate (although comparable to FFT, which is also a spectral technique).

In Figure 10.6, we show that starting with two different initial profiles, namely, one in which the change from zero to unity is abrupt and one in which the change from zero to unity is ramped, we obtain the same solution using the wavelet-collocation approach. Finally, in Figure 10.7, we show that

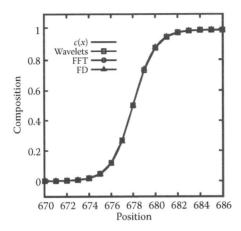

FIGURE 10.5
The comparison of the composition profile obtained using finite difference (FD), FFT, and wavelet-collocation implementations for the 1-D Cahn–Hilliard equation with the analytical solution (which is shown by the continuous line). Note that for the FFT solution, the system size is twice as large as the FD and wavelet-collocation schemes, and that there are two interfaces, of which, only one is shown. The calculations are done on a 1×1024 grid. For clarity, we show only the region near the interface, corresponding to the range 660 to 700.

using different κ values, we obtain different interfacial widths for the equilibrium profile. In Table 10.3, we show the values of interfacial energy and width calculated for different values of κ; we see that these numbers scale as expected, that is, as $\sqrt{\kappa}$.

It is also possible to solve the Cahn–Hilliard equation (i.e., Equation 10.9) for different initial and boundary conditions, namely, a uniform composition of 0.5 (with very small noise) and periodic boundary conditions. In such a case, by decomposing the composition fluctuations into the corresponding Fourier components and following their time evolution, for the very early

TABLE 10.2

L^2 Norm Errors for Three Different Implementations

Initial Profile	κ	FD	FFT	Wavelet
Ramped	1	2.0146e-05	7.39989e-07	7.30938e-08
Ramped	2	0.00233101	0.00258896	0.00254043
Ramped	4	0.00768298	0.00783656	0.00785316
Ramped	8	0.0139625	0.0147265	0.0149402
Step	1	0.281004	0.279586	0.0662797

Note: The L^2 norm errors in solution as compared with the analytical solution for different initial profiles (ramped and step) and κ different parameters obtained using different implementations: finite difference (FD), Fourier spectral (FFT), and wavelet-collocation (wavelet).

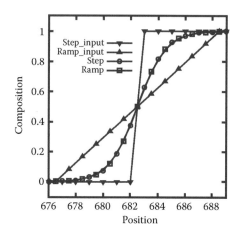

FIGURE 10.6

The comparison of the composition profile obtained using wavelet-collocation implementations for the 1-D Cahn–Hilliard equation using two different initial conditions; the solution obtained using the ramped initial profile is shown by a continuous line and that obtained with a step profile is shown by points. In these two cases, the interface is not formed at the same site; so we have overlaid the solutions by shifting the solutions to match the point at which the value of 0.5 in the composition is reached. The calculations are done on a 1 × 1024 grid. For clarity, we show only the region near the interface, corresponding to the range 660 to 700.

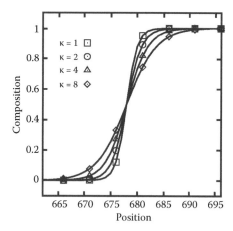

FIGURE 10.7

The composition profiles obtained using wavelet-collocation implementations for the 1-D Cahn–Hilliard equation using different κ values; as expected, with increasing κ values, the interface width also increases. The calculations are done on a 1 × 1024 grid. For clarity, we show only the region near the interface, corresponding to the range 660 to 700.

TABLE 10.3

Values of Interfacial Energies and Widths

κ	Interfacial Width	Interfacial Energy
1	8.164385	0.333320
2	11.431422	0.471404
4	16.084023	0.666665
8	22.690719	0.942801

Note: The interfacial widths and energies calculated from the 1-D profile obtained using the wavelet implementation; the scaling behavior of these quantities are as expected.

stages of decomposition, with wavenumbers which are less than a critical wavenumber will grow and those above the critical wavenumber will decay (Hilliard, 1970). This is because the gain in free energy due to decomposition is less than the increase in interfacial energy due to the creation incipient interfaces for larger wave numbers (which correspond to smaller wavelengths). Using the given bulk free energy density and gradient energy coefficient, one can further show that the critical wave number in a system in which we do not consider elastic stress effects (see Appendix A) is given by the following expression:

$$\beta_c = \sqrt{\frac{-f''}{2\kappa}} \tag{10.49}$$

where f'' is the second derivative of the bulk free energy density with respect to composition. In Figure 10.8, we show the plot of amplification factor (normalized rate of growth of Fourier components, see Appendix A) as a function of their wave numbers in a simulation carried out on a system of 1024 nodes with $\kappa = A = M = 1$ and mean composition 0.5. The plot corresponds to time $t = 10$. The $\Delta x = 1$ and $\Delta t = 0.001$; furthermore, the initial noise is $\pm 0.5 \times 10^{-4}$. From this plot, we calculate a critical wave number of 115.244, which agrees very well with the analytical value of 115.240 calculated using Equation 10.49 and the used input parameters. It was also observed from the plot that the value of maximally growing wave number (β_{max}) 81.489 agrees well with the analytical expression $\beta_{max} = \frac{\beta_c}{\sqrt{2}} = 81.487$. These results indicate that in addition to getting the time-independent profile correctly, the wavelet-based scheme also captures the dynamics of the transformation quite well.

We are working on wavelet implementations in two dimensions and three dimensions as well as on the use of continuous wavelets with compact support. The results from these studies will be reported elsewhere.

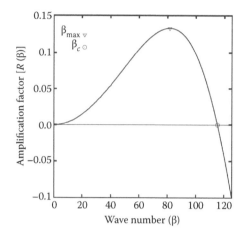

FIGURE 10.8

Amplification factor $R(\beta)$ is plotted against β. β_c can be found at intersection of $R(\beta)$ and x axis. β_{max} is a point for which $\dfrac{dR(\beta)}{d\beta} = 0$. At the very early stages of decomposition, β_c and β_{max} are independent of time.

Conclusions

Phase transitions and the resultant microstructures play an important role in understanding the dynamics of the compositions within the Earth's crust, mantle, and core, particularly at the boundaries that distinguish these layers. In this chapter, we have made an attempt to show how phase field models can be used to study such phase transition dynamics and the formation of microstructures. Specifically, we have shown that the elastic stress-driven ATG instability in a mineral in contact with its solution, which is of interest in the study of the mechanical behavior within the Earth's deep mantle, can be successfully modeled using a phase field model. We have used Fourier spectral technique for the implementation of the phase field model. Both the Cahn–Hilliard equation and the equation of mechanical equilibrium are solved accurately and efficiently using this technique. We have shown that whereas imposing the periodic boundary conditions is a major constraint in the use of Fourier spectral technique for solving the Cahn–Hilliard equation, the same can be easily overcome by using the wavelet-collocation implementation of the Cahn–Hilliard equation and thus the nonperiodic boundary conditions can also be imposed in the solution of the problem. We have solved the 1-D Cahn–Hilliard equation using the wavelet-collocation approach and benchmarked the 1-D solutions using both analytical solutions and numerical solutions obtained using finite difference and Fourier spectral techniques. Our preliminary results show that the wavelet-based solution for phase field models is promising and we have highlighted how both phase field models and their wavelet implementations will be of much interest to the geoscience community.

Appendix A

Consider the Cahn–Hilliard equation (Equation 10.33):

$$\frac{\partial c}{\partial t} = M \frac{\partial^2}{\partial x^2}\left(h - 2\kappa \frac{\partial^2 c}{\partial x^2} \right) \tag{10.A1}$$

If the gradient energy coefficient κ is assumed to be independent of composition (and hence position), we can write the Cahn–Hilliard equation as

$$\frac{\partial c}{\partial t} = M\left[\frac{\partial^2 f}{\partial c^2} - 2\kappa \frac{\partial^2}{\partial x^2} \right]\frac{\partial^2}{\partial x^2} \tag{10.A2}$$

Let us also assume that the spatial variation of the composition can be described by the following Fourier integral:

$$c = \int A(\beta)\exp(i\beta x)\,d\beta \tag{10.A3}$$

where $A(\beta)$ is the amplitude of Fourier component of wave number β. Substituting Equation 10.A3 in Equation 10.A2, we get

$$\frac{dA(\beta)}{dt} = -M[f'' + 2\kappa\beta^2]\beta^2 A(\beta) \tag{10.A4}$$

where f'' is $\dfrac{\partial^2 f}{\partial c^2}$. This ordinary differential Equation 10.A4 can be solved to obtain the solution

$$A(\beta, t) = A(\beta, 0)\exp[R(\beta)t] \tag{10.A5}$$

where $R(\beta)$ is the amplification factor given by

$$R(\beta) = -M[f'' + 2\kappa\beta^2]\beta^2 \tag{10.A6}$$

If $R(\beta)$ is negative, then the corresponding component in the Fourier decomposition decays, whereas if it is positive then the corresponding component grows. Hence, we can find the critical wave number β_c, which neither grows nor decays by equating to zero (Equation 10.A6) and replacing β with β_c. Hence, we get

$$\beta_c = \sqrt{\frac{-f''}{2\kappa}} \qquad (10.A7)$$

References

Asaro, R.J., and Tiller, W.A. 1972. Interface morphology development during stress corrosion cracking: Part I. Via surface diffusion. *Metallurgical Transactions* 3, 1789–1796.

Bahr, K., Olsen, N., and Shankland, T.J. 1993. On the combination of the magnetotelluric and the geomagnetic depth sounding method for resolving an electrical conductivity increase at 400 km depth. *Geophysical Research Letters* 20, 2937–2940.

Berrehar, J., Caroli, C., Lapersonne-Meyer, C., and Schott, M. 1992. Surface patterns on single-crystal films under uniaxial stress: Experimental evidence for the Grinfeld instability. *Physical Review B* 46, 13487–13495.

Beylkin, G. 1992. On the representation of operators in bases of compactly supported wavelets. *SIAM Journal on Numerical Analysis* 29, 1716–1740.

Boettinger, W.J., Warren, J.A., Beckermann, C., and Karma, A. 2002. Phase-field simulation of solidification. *Annual Review of Materials Research* 32, 163–194.

Chandrasekhar, E. 1999. Determination of Deep Earth Electrical Conductivity Using Long Period Geomagnetic Variations. PhD Thesis, University of Mumbai, Mumbai, India, 201 pp.

Chandrasekhar, E. 2000. Geo-electrical structure of the mantle beneath the Indian region derived from the 27-day variation and its harmonics. *Earth, Planets and Space* 52, 587–594.

Chen, L.-Q. 2002. Phase-field models for microstructural evolution. *Annual Review of Materials Research* 32, 113–140.

Chirranjeevi, B.G., Abinandanan, T.A., and Gururajan, M.P. 2009. A phase field study of morphological instabilities in multilayer thin films. *Acta Materialia* 57, 1060–1067.

Cogswell, D.A. 2010. A Phase-Field Study of Ternary Multiphase Microstructures. PhD Thesis, Massachusetts Institute of Technology, Cambridge, MA. 179 pp.

Constable, S.C. 1993. Constraints on mantle electrical conductivity from field and laboratory measurements. *Journal of Geomagnetism and Geoelectricity* 45, 1–22.

Daubechies, I. 1992. *Ten Lectures on Wavelets*. CBMS-NSF Series in Applied Mathematics, SIAM, Philadelphia.

den Brok, B., Morel, J., and Zahid, M. 2002. In situ experimental study of roughness development at a stressed solid/fluid interface. In *Deformation Mechanisms, Rheology and Tectonics: Current Status and Future Perspectives*, edited by De Meer, S., Drury, M.R., de Bresser, J.H.P., and Pennock, G.M., 200, 73–83. Geological Society, London, Special Publications.

Gopalakrishnan, S., and Mira, M. 2010. *Wavelet Methods for Dynamical Problems.* CRC Press, Boca Raton, FL.

Grinfeld, M.A. 1986. Instability of the interface between a non-hydrostatically stressed elastic body and melts. *Soviet Physics-Doklady* 290, 1358–1363.

Gururajan, M.P. 2006. Elastic Inhomogeneity Effects on Microstructures: A Phase Field Study. PhD Thesis, Indian Institute of Science, Bangalore.

Gururajan, M.P., and Abinandanan, T.A. 2007. Phase field study of precipitate rafting under a uniaxial stress. *Acta Materialia* 55, 5015–5026.

Gutenberg, B. 1948. On the layer of relatively low wave velocity at a depth of about 80 km. *Bulletin of the Seismological Society of America* 38, 121–148.

Helffrich, G., and Wood, B. 1996. 410 km discontinuity sharpness and the form of the olivine α-β phase diagram: Resolution of apparent seismic contradictions. *Geophysical Journal International* 126, F7–F12.

Hilliard, J.E. 1970. Spinodal decomposition. In *Phase Transformations.* American Society for Metals, Metals Park, Ohio.

Hu, S.Y., and Chen, L.Q. 2001. A phase-field model for evolving microstructures with strong elastic inhomogeneity. *Acta Materialia* 49, 1879–1890.

Jefferys, H. 1939. The times of the core waves. *Monthly Notices of the Royal Astronomical Society. Geophysical Supplement* 4, 498–533.

Khachaturyan, A.G., Semenovskaya, S., and Tsakalakos, T. 1995. Elastic strain energy of inhomogeneous solids. *Physical Review B* 52(22), 15909–15919.

Koehn, D., Arnold, J., Jamtveit, B., and Malthe-Sorenssen, A. 2003. Instabilities in stress corrosion and the transition to brittle failure. *American Journal of Science* 303, 956–971.

Michel, J.C., Moulinec, H., and Suquet, P. 1999. Effective properties of composite materials with periodic microstructure: A computational approach. *Computer Methods in Applied Mechanics and Engineering* 172, 109–143.

Morel, J., and den Brok, S.W.J. 2001. Increase in dissolution rate of sodium chlorate induced by elastic strain. *Journal of Crystal Growth* 222, 637–644.

Ohtani, E., and Sakai, T. 2008. Recent advances in the study of mantle phase transitions. *Physics of the Earth and Planetary Interiors* 170, 240–247

Olsen, N. 1998. The electrical conductivity of the mantle beneath Europe derived from C- responses from 3 to 720 hours. *Geophysical Journal International* 133, 298–308.

Omura, K. 1991. Change of electrical conductivity of olivine associated with olivine–spinel transition. *Physics of the Earth and Planetary Interiors* 65, 292–307.

Pidduck, A.J., Robbins, D.J., Cullis, A.G., Leong, W.Y., and Pitt, A.M. 1992. Evolution of surface morphology and strain during SiGe epitaxy. *Thin Solid Films* 222, 78–84.

Torii, R.H., and Balibar, S. 1992. Helium crystals under stress: The Grinfeld instability. *Journal of Low Temperature Physics* 89, 391–400.

Xu, Y., Poe, B.T., Shankland, T.J., and Rubie, D.C. 1998. Electrical conductivity of olivine, wadsleyite, and ringwoodite under upper-mantle conditions. *Science* 280, 1415.

Wang, D., and Pan, J. 2004. A wavelet-Galerkin scheme for the phase field model of microstructural evolution of materials. *Computational Materials Science* 29, 221–242.

Williams, J.R., and Amaratunga, K. 1997. A discrete wavelet transform without edge effects using wavelet extrapolation. *Journal of Fourier Analysis and Applications* 3, 435–449.

Index

Page numbers followed by f and t indicate figures and tables, respectively.

A

A-9/7, filter banks design approach, 81

ABG (Alibag) observatory, GJ study and, 206t, 209, 210f, 213, 215

Abisko (ABK) observatory, GJ study and, 204, 205t, 214f

Admissibility condition
  defined, 52
  in wavelet functions, 5, 6

Akalagavi cave, 220
  multitaper analysis of, 222f

Alert (ALE) observatory, GJ study and, 204, 205t, 214f

Alias system function, 41

Alibag (ABG) observatory, GJ study and, 206t, 209, 210f, 213, 215

Analysis equations, defined, 36

Analysis filter bank, 39, 39f

Analysis modulation matrix, defined, 42

Analyzing wavelet (mother wavelet), defined, 6

Apia (API), GJ study and, 206t, 209, 211f, 213, 214f, 215f

Approximate coefficients, in DWT, 10

Approximation subspaces, defined, 54

Asaro–Tiller–Grinfeld (ATG) instabilities
  defined, 251
  equations for, 251–255, 251f
    chemical free energy, 252, 253, 254
    elastic energy, 252–253, 254
    order parameters in, 251–253
    in He-IV solid–liquid interfaces, 251
  solving equation of mechanical equilibrium, 254–255

ATG. See Asaro–Tiller–Grinfeld (ATG) instabilities

B

B-9/7, filter banks design approach, 81

Baker Lake (BLC) observatory, GJ study and, 204, 205t

Bangui (BNG) observatory, GJ study and, 206t, 209, 210f, 214f, 215f

Benchmarking, 1-D Cahn–Hilliard equation and, 259–268

Biorthogonal multiresolution analysis, 55–57

Biorthogonal wavelet filter banks (BWFBs), 69–70, 80, 81, 82t

BLC (Baker Lake) observatory, GJ study and, 204, 205t

Block transform (BT), 30

BNG (Bangui) observatory, GJ study and, 206t, 209, 210f, 214f, 215f

Bouger gravity anomaly, 112–113

Boundaries, phase field modeling evolution of solid–liquid and solid–solid. See Phase field modeling

Box-counting dimension, defined, 131

Box counting method
  for fractal dimension analysis, 155, 159–161, 160f, 161f, 162f
  fractal dimension estimation, 14–15
  multifractal analysis in earthquake prediction, 182, 185

BWFBs (biorthogonal wavelet filter banks), 69–70, 80, 81, 82t

C

Cahn–Hilliard (1-D nonlinear diffusion) equation, 248, 261–264, 269–270
  results from, 264–267, 265f, 265t, 266f, 267t, 268f

Cantor set, 178
  fractal object, 132, 132f

Capacity dimension, defined, 179

Cascade algorithm, generating scaling
        functions and wavelets, 60–63,
        63f, 64f, 65f
CCWT (complex continuous wavelet
        transform), 204
CDF (Cohen–Daubechies–Feauveau)
        wavelet family, 73
Chemical free energy, in ATG
        instabilities, 252, 253, 254
CMO (College) observatory, GJ study
        and, 204, 205t, 207, 207f, 213,
        214f
Cohen–Daubechies–Feauveau (CDF)
        wavelet family, 73
College (CMO) observatory, GJ study
        and, 204, 205t, 207, 207f, 213,
        214f
Compactly supported wavelets
    Daubechies wavelets, 260
    functions, 5
    orthogonal wavelets design
        on interval [0, 1], 72
        on interval [0, 3], 72–73
Compact support, wavelet properties, 7
Complex continuous wavelet transform
        (CCWT), 204
Complex wavelets
    analysis of GJs, 204
    in geomagnetism, 199. *See also*
        Geomagnetic jerks (GJs)
Constant relative bandwidth filter bank
        (constant-Q), 52
Continuous-time signal, uncertainty
        principle, 76–78
Continuous wavelet transform (CWT),
        6–7, 98–99
    coefficient calculation, 8, 9f
    DWT *vs.*, 52–53
    GJ events and, 201–202
    nonstationary behavior in
        temperature anomaly and tree-
        ring data and, 230
    rainfall anomalies, 235–236, 236f,
        237f
    uses in time series analysis, 233
    time shift in, 7–8
Contours, defined, 148
Convolution matrix, defined, 64
Correlation dimension, defined, 180

Correlation integral method, fractal
        dimensions estimation, 15–16
Cosine functions, in Fourier theory, 1–2
Critically sampled (maximally
        decimated) filter banks, 39
Crust–mantle boundary, 248
CWT. *See* Continuous wavelet transform
        (CWT)

D

Data compression, wavelet-based
    multiscale processing for, 118,
        123–131
        algorithm, 128–131, 129f, 130f, 131f
        coarser scale, 127
        data decomposition, 124, 125, 125f
        DWT of Lena image, example,
            126f
        encoders, 124, 125–126
        example, 125–128, 125f, 126f, 127f
        EZW, 118, 125
        finer scale, 127
        LL1 subband, 126
        LL2 subband, 127
        spatial-orientation tree, 127–128,
            127f
        SPIHT algorithm, 118, 125, 126–127
Daubechies, Ingrid, 97
Daubechies wavelets
    Cahn–Hilliard equations and,
        261–264
    compactly supported, 260
DCWT (discretized version of CWT),
        99–100
Decomposition of data, in data
        compression, 124, 125, 125f
Demodulates, 3–4
    defined, 3
Denoising, 112
    data using DWT, rainfall anomalies
        and, 234
    signal, thresholding techniques for,
        141–143, 142f, 144f
    wavelets in, 141
Depth estimation, wavelet-based
    gravity data for, 170–172, 171f, 172f
Detailed coefficients, in DWT, 10
Detail subspaces, defined, 54

Detectability limit, of geophysical survey network, 166–167, 167t
Detrending, tree-ring width data, 240–241
Diffuse interface models. *See* Phase field modeling
Dilation equation, for dyadic scaling function, 54
Discrete cosine transform (DCT), 124
Discrete-time signal, uncertainty principle
  direct extension from continuous-time measure, 78–79
Discrete wavelet transform (DWT), 9–11, 10f, 99–102, 118
  approximation coefficients, 10, 101–102
  CWT *vs.*, 52–53
  detailed coefficients, 10, 100–102
  dyadic wavelet transformation, 9
  GJ events and, 201–202
  nonstationary behavior in
    temperature anomaly and tree-ring data, 229–230
    denoising data, rainfall anomalies and, 234
    through Haar wavelets, 231–233
  process, 10–11, 10f
  scaling functions, 100–101
  shift-invariant property for, 8
  subband coding, 100
  wavelet functions, 100
Discretized version of CWT (DCWT), 99–100
Double shift orthogonality conditions, 45
$D_q – q$ spectra, in earthquake prediction, 188, 189, 190f
D_q spectra, in earthquake prediction, 183, 185, 186, 188, 189
DWT. *See* Discrete wavelet transform (DWT)
Dyadic MRA, 118, 119–124
  axioms of, 119–122
    in analyzing self-similar functions, 121
    intersection of all subspaces is trivial subspace , 121
    ladder axiom, 119–121, 120f

  orthogonal basis, 122
  perfect reconstruction, 121
Dyadic sampling, 53
Dyadic wavelet transformation. *See also* Discrete wavelet transform (DWT)
  defined, 9

E

Earth
  permanent magnetic field (dipole field) of, 195–196
  defined, 195
  phase transitions in, 247–248, 249f
Earthquake prediction, multifractal studies in. *See* Multifractals and multifractal analysis, in earthquake prediction
ECG (electrocardiogram) signals, one-dimensional singularity in
singularities detection (case study), 143, 144f, 145–148, 145f, 146f, 147f
Elastic energy, in ATG instabilities, 252–253, 254
Electrocardiogram (ECG) signals, one-dimensional singularity in
singularities detection (case study), 143, 144f, 145–148, 145f, 146f, 147f
Embedded zero wavelet (EZW), 118, 125
Encoders, in data compression, 124, 125–126
Epicenters (hypocenters) of earthquakes
  densities of fractures/faults and, 181
  distribution of, 181, 187, 189
  D_q spectra and, 186
  Muzaffrabad–Kashmir earthquake, 184, 188
Equation of mechanical equilibrium
  Fourier spectral technique and, 255, 256, 257
  solution, in ATG instabilities, 254–255
Eskdalemuir (ESK) observatory, GJ study and, 205t, 208f, 209, 212, 214f
EZW (embedded zero wavelet), 118, 125

F

Fast wavelet transform (FWT), 57–59, 59f, 103–104
Father wavelet, 54
　defined, 7
FGW (first-generation wavelets)
　applications for generalizations in, 108
　properties, 107
Fidelity factor, defined, 4
Filter banks, 31, 38–52
　design, time–frequency localization, 79–80
　design methodology, 80–90, 82t, 83t, 84f, 85f, 85t, 86t, 87t, 88f, 89f, 90f
　　results and comparisons, 81–90, 82t, 83t, 84f, 85f, 85t, 86t, 87t, 88f, 89f, 90f
　for Haar wavelet, 36–37, 37f
　linear phase biorthogonal, design parametrization technique, 73–76
　octave, 51–52, 51f
　perfect reconstruction (PR). *See* Perfect reconstruction (PR) filter banks
　wavelet, design, 69–71
　wavelets and, connection, 54–69
　　condition for convergence in $L_2(R)$, 63–67, 65f, 67f
　　FWT, 57–59, 59f
　　generating scaling functions and, 60–63, 63f, 64f, 65f
　　MRA. *See* Multiresolution analysis (MRA)
　　PR filter banks and, 59–60
　　regularity measure, 68–69
　　transition matrix, 64–67, 65f, 67f
Filtering steps, in 331-year-long proxy monsoon data, 222–225, 223f, 224f, 224t, 225f, 226f, 227f
Finite impulse response (FIR) filters, 41, 42, 44
FIR (finite impulse response) filters, 41, 42, 44
First-generation wavelets (FGW)
　applications for generalizations in, 108
　properties, 107

Fixed-mass method
　fractal dimensions estimation, 15
　multifractal analysis in earthquake prediction, 182, 183, 185
Fixed-radius method, multifractal analysis in earthquake prediction, 182, 183, 185
Fixed-size algorithm, fractal dimensions estimation, 15
Fourier, Jean Baptiste Joseph, 1
Fourier transform (FT), 1–2, 30, 93–94
　based spectral technique, for implementation of phase field model, 248, 255–259. *See also* Phase field modeling
　　nondimensionalization, 257
　　results, 257–259, 258f, 258t, 259f
　drawback of, 3, 30
　rainfall data using WT, analysis, 237–238, 238f
　spectra, 2f
　STFT. *See* Short-time Fourier transform (STFT)
　useful features of signal and, 30
　WFT, 3
　WT, 30–31
Fractal (fractional) dimension
　capacity dimension, 179
　correlation dimension, 180
　defined, 178
　determination
　　box counting method for, 155, 159–161, 160f, 161f, 162f
　　survey network design, 163, 164f. *See also* Survey network design
　　estimation, 177, 178, 179–180
　　power law in, 177, 179
　information dimension, 180
　similarity dimension, 179
Fractal dimension, multifractal analysis in earthquake prediction, 181, 182, 183
Fractal form, defined, 178
Fractals, 131–133
　application, 18
　in applied geophysics, 155–173
　　box counting method, for fractal dimension. *See* Box counting method

fractal time series
  characterization. *See* Fractal
  time series characterization
  overview, 155–156
  R/S analysis, 157–159, 158f
  survey network design. *See*
  Survey network design
Cantor set, 132, 132f
defined, 118, 131
dimensions estimation, methods,
  13–16
  box counting method, 14–15
  correlation integral method, 15–16
  fixed-mass algorithm, 15
  fixed-size algorithm, 15
in geological context
  Cantor set, 178
  fractal dimension, 179–180
  in geological context, 177–178
  Koch curve, 178
  multifractals and multifractal
    analysis. *See* Multifractals and
    multifractal analysis
  similarity and scaling, 178–179
in geophysics, 18
IFS, 133
Koch curve, 132–133, 132f
overview, 4–5
self-similarity in. *See* Self-similar
  functions
in time series analysis, 11–16, 12f
  density, of boreholes, 11–12, 12f,
    13f
  persistence of, 11
  reflectivity sequence, of boreholes,
    11–12, 12f, 13f
  scaling spectral method, 13
  spectral analysis method, 12–13
  susceptibility distribution, of
    boreholes, 11–12, 12f, 13f
wavelets and, 18–20
  wavelet-based fractal analysis, 20
Fractal time series characterization
  fBm concept. *See* Fractional
    Brownian motion (fBm)
  fGn concept. *See* Fractional Gaussian
    noise (fGn)
  Hurst coefficient in, 156, 157
  R/S analysis, 157–159, 158f

Fractional Brownian motion (fBm),
  118–119, 134, 155, 156–157
  Hurst parameters in, 136–137
Fractional Gaussian noise (fGn), 155,
  156–157
Frequency center, defined, 76
Frequency spread (frequency variance),
  defined, 76
FWT (fast wavelet transform), 57–59, 59f,
  103–104

G

Gabor wavelet, 204
Geomagnetic jerks (GJs), 195–216
  data processing of global magnetic
    observatory data, 199–200, 200f,
    201f, 205t–206t
  defined, 17, 196
  discussion, 209, 212–216, 214f, 215f
  origin, 197
  overview, 195–198, 196f, 197f
  region-wise study, 204–209
    high-latitude to mid-latitude
      region, 205t–206t, 207–209, 208f
    mid-latitude to equatorial region,
      205t–206t, 209, 210f
    polar region, 204–207, 205t–206t,
      207f
    southern hemisphere region,
      205t–206t, 209, 211f
  wavelet analysis, 201–204
    complex wavelet analysis of, 204
    CWT, 201–202
    DWT, 201–202
    estimation of ridge functions
      using real wavelets, 202–203,
      203f, 212, 213, 216
  wavelets and complex wavelets in
    geomagnetism, 199
Geomagnetic secular variations,
  defined, 196
Geomagnetism, wavelets and complex
  wavelets in, 199. *See also*
  Geomagnetic jerks (GJs)
Geometry, fractal
  multifractal studies in earthquake
    prediction. *See* Multifractals
    and multifractal analysis

Geophysics
  applied
    fractals in. *See* Fractals, in applied geophysics
    wavelets in. *See* Wavelets, in applied geophysics
  fractals in, 18
  wavelets in, 16–17
GJs. *See* Geomagnetic jerks (GJs)
Global magnetic observatory data
  data processing, 199–200, 200f, 201f, 205t–206t
Gradient-based inversion, matrix multiplication for
  in geophysical data analysis, 167–170, 169f
Gravity data, WT
  for source depth estimation, 170–172, 171f, 172f
Grossman, Alex, 97
Gutenberg discontinuity. *See* Mantle–core boundary

H

Haar, Alfred, 97
Haar filter bank, 36–37, 37f
Haar wavelets, 97
  construction, 32–38
    analysis equations, 36
    filter bank for, 36–37, 37f
    Haar scaling function, 33, 33f, 36
    Haar wavelet function, 35, 35f, 36
    MRA, 36
    PCA and. *See* Piecewise constant approximation (PCA)
    synthesis equations, 36
  DWT through, 231–233
  scaling function for, 19, 19f
Half-band filter, 44
Hausdorff–Bescovitch dimension. *See* Fractal (fractional) dimension
Heisenberg uncertainty principle, 76, 95–96
He-IV solid–liquid interfaces, ATG instabilities in, 251
Hermanus (HER), GJ study and, 206t, 209, 211f, 213, 214f, 215f

Heterogeneous fractals, earthquake prediction and, 181–182, 183
Hidden cycles, during wavelet analysis paleomonsoon data from speleothems and, 222–225, 223f, 224f, 224t, 225f, 226f, 227f
High-latitude to mid-latitude region GJ study in, 205t–206t, 207–209, 208f
Himalaya region, seismicity of, 183, 184, 185f, 186, 188, 189
Holder regularity measure, 68
Homogeneous fractals, earthquake prediction and, 182–183
Hooke's law of elasticity, 253, 254
Hurst coefficient, in time series characterization, 156, 157
  R/S analysis, 157–159, 158f
Hurst exponent, defined, 20
Hurst parameters
  defined, 136
  in fractional Brownian motion, 136–137
Hypocenters (epicenters) of earthquakes
  densities of fractures/faults and, 181
  distribution of, 181, 187, 189
  D_q spectra and, 186
  Muzaffrabad–Kashmir earthquake, 184, 188

I

IAGA (International Association of Geomagnetism and Aeronomy) codes, 200f, 205t–206t
IFS (iterative function scheme), 133
IIR (infinite impulse response) filters, 42
Infinite impulse response (IIR) filters, 42
Information dimension, defined, 180
International Association of Geomagnetism and Aeronomy (IAGA) codes, 200f, 205t–206t
International Tree-Ring Data Bank, 240
Inversion, gradient-based
  matrix multiplication, in geophysical data analysis, 167–170, 169f
Iterated function system (IFS), 19, 133

K

Kakioka (KAK) observatory, GJ study and, 205t, 208f, 209, 213, 214f, 215
Kashmir–Muzaffrabad earthquake, 184, 187–188, 189, 190f
Koch curve, fractal object, 132–133, 132f

L

Ladder axiom, of dyadic MRA, 119–121, 120f
Lagrange half-band polynomial (LHBP), 49
  factorization of, 69–70
Lattice structures, wavelet filter banks design method, 70
Lawton matrix. *See* Transition matrix
Leirvogur (LRV) observatory, GJ study and, 204, 205t, 214f
Lerwick (LER) observatory, GJ study and, 200, 201f, 204, 205t, 207, 207f, 213, 214f
LHBP (Lagrange half-band polynomial), 49
  factorization of, 69–70
Lifting scheme, 109
  defined, 108
  wavelet filter banks design method, 70
Lifting step, defined, 109
Linear phase biorthogonal filter banks, 47–48
  design, parametrization technique, 73–76
Logarithmic filter bank, 52
Low-pass filters, for smoothing noisy data, 223
Low power, extracting
  during wavelet analysis, paleomonsoon data from speleothems, 222–225, 223f, 224f, 224t, 225f, 226f, 227f
LRV (Leirvogur) observatory, GJ study and, 204, 205t, 214f

M

Magnetic observatory data, global data processing, 199–200, 200f, 201f, 205t–206t

Mallat, Stephane, 57, 97, 133, 139
Mandelbrot, Benoit B., 4–5
Mantle–core boundary, 248
Mathematical microscopes, wavelets, 8
MATLAB, 81, 225, 234
  coding
    CWT coefficients, 235, 236
    FFT in, 238
    for matrix multiplication, 168–170
    in time series data of rainfall anomalies, 234–235
Matrix multiplication, for gradient-based inversion
  in geophysical data analysis, 167–170, 169f
Maxflat half-band filter, 48–50, 50f, 51t
Maximally decimated (critically sampled) filter banks, 39
  two-channel, 40, 40f
MBour (MBO) observatory, GJ study and, 206t, 209, 210f, 214f
Mechanical equilibrium, equation of solution, in ATG instabilities, 254–255
Meyer, Yves, 53, 97
MF-DFA (multifractal–detrended fluctuation analysis), 184
Mid-latitude to equatorial region
  GJ study in, 205t–206t, 209, 210f
Mohorovicic discontinuity, 248
Monsoon, paleomonsoon data from speleothems
  WT application. *See* Paleomonsoon data from speleothems
Morlet, Jean, 16, 97
Morlet wavelet, defined, 233
Mother wavelet (analyzing wavelet)
  defined, 6, 52
  translated and scaled, 97
MRA. *See* Multiresolution analysis (MRA)
Multifractal–detrended fluctuation analysis (MF-DFA), 184
Multifractals and multifractal analysis, in earthquake prediction, 180–184
  box counting method, 182, 185
  $D_q - q$ spectra, 188, 189, 190f
  D_q spectra, 183, 185, 186, 188, 189

epicenters (hypocenters), 181, 184,
      186, 187, 188, 189
   fixed-mass method, 182, 183, 185
   fixed-radius method, 182, 183, 185
   fractal dimension in, 181, 182, 183
   heterogeneous fractals, 181–182, 183
   homogeneous fractals, 182–183
   Muzaffrabad–Kashmir earthquake,
      184, 187–188, 189, 190f
   scaling property, 181, 182
   of seismicity, 184–190, 185f, 186f, 187f,
      188t, 189f, 190f
   seismotectonic map, 181
   self-similar, 181
   southern Italy (case study), 184
Multiresolution analysis (MRA), 36, 54–57
   biorthogonal, 55–57
   dyadic. *See* Dyadic MRA
   theorem, 122–125
Multiscale processing, 118–153
   data compression. *See* Data
      compression, wavelet-based
   fractals, 118, 131–133
   MRA
      dyadic. *See* Dyadic MRA
      theorem, 122–125
   overview, 118–119
   self-similar functions. *See* Self-
      similar functions
   singularities. *See* Singularities
Multitaper method, 220–222, 221f, 222f
Muzaffrabad–Kashmir earthquake, 184,
      187–188, 189, 190f

N

National Climatic Data Center (NCDC),
      234
National Oceanic and Atmospheric
      Administration (NOAA), 234,
      240
NCDC (National Climatic Data Center),
      234
Niemegk (NGK) observatory, GJ study
      and, 205t, 213
9/7, filter banks, 80–81
NOAA (National Oceanic and
      Atmospheric Administration),
      234, 240

Noise removal, singularities and, 119, 138
   detection
      denoising. *See* Denoising
      noisy signals, 140–141, 140f
      signal without noise, 139, 139f
Nondimensionalization, in phase field
      modeling, 257
Nonstationary behavior, in temperature
      anomaly and tree- ring data
   wavelet perspective, 229–243
      analysis of rainfall data using. *See*
         Rainfall data using WT
      applications, 234–243
      CWT, 230, 233
      DWT, 229–230, 231–233
      extracting wavelet coefficients
         and fitting on time series plot,
         239, 239f, 240f
      overview, 229–230
      STFT, 230
      tree-ring width data using. *See*
         Tree-ring width data using WT
      WT, 230–231, 234–238

O

Observatories, global magnetic. *See also*
      specific entries
   data processing, 199–200, 200f, 201f,
      205t–206t
Octave filter banks, 51–52, 51f
   logarithmic, 52
   tree structured, 51, 51f
One-dimensional singularity in ECG
      signals
   singularities detection (case study),
      143, 144f, 145–148, 145f, 146f, 147f
1-D nonlinear diffusion (Cahn–Hilliard)
      equation, 261–264, 269–270
   results from, 264–267, 265f, 265t, 266f,
      267t, 268f
Optimum grid, for survey network
      design. *See also* Survey network
      design
   governing factors, 164–165, 165f, 166f
Order parameters, in phase field
      models, 250, 251–253
Orthogonal basis axioms, in dyadic
      MRA, 122

Orthogonal matrix, defined, 45
Orthogonal wavelets design, 71–73
  compactly supported
    on interval [0, 1], 72
    on interval [0, 3], 72–73

P

Paleoclimate studies, 17
Paleomonsoon data from speleothems
  WT application, 219–227. *See also*
    Speleothems
    extracting low power, hidden
      cycles during wavelet analysis,
      222–225, 223f, 224f, 224t, 225f,
      226f, 227f
    multitaper analysis, 220–222, 221f,
      222f
    overview, 219–220
Pamatai (PPT), GJ study and, 206t, 209,
    211f, 213, 214f, 215f
Parametrization
  of filter coefficients, wavelet filter
    banks design method, 70
  technique, linear phase biorthogonal
    filter banks design, 73–76
Paraunitary filter banks, 44–47
  double shift orthogonality
    conditions, 45
  example, 46–47
  important features, 45–46
  orthogonal, 45
  polynomial, 45
PCA (piecewise constant
    approximation), Haar wavelets
    and, 32–38
  box function in, 33, 33f
  Haar filter bank and, 37
  Haar wavelet function, 35, 35f, 36
  on half interval, 34–35, 34f, 35f, 37
  on unit interval, 32, 32f, 37
Perfect reconstruction, axioms of
  in dyadic MRA, 121
Perfect reconstruction (PR) filter banks,
    38–51
  analysis bank, 39, 39f
    SBC, 39
    subband signals in, 39
  decimator and interpolator, 38, 38f

design, product filter and, 43–44
  half-band filter, 44
  linear phase biorthogonal, 47–48
  maxflat half-band, 48–50, 50f, 51t
  maximally decimated (critically
    sampled), 39
  paraunitary. *See* Paraunitary filter
    banks
  synthesis bank, 39, 39f
  two-channel. *See* Two-channel PR
    filter banks
  wavelet filter banks and, 59–60
Persistence, of time series, fractals in, 11
PF (potential fields), applications of WT
    to, 110–114
Phase field modeling, of evolution of
    solid–solid and solid–liquid
    boundaries, 247–270
  advantage, 250
  evolution equations and their
    derivation, 251–255
    ATG instabilities. *See* Asaro–
      Tiller–Grinfeld (ATG)
      instabilities
  order parameters in, 250
  overview, 250–255
  phase transitions in earth,
    understanding of, 247–248, 249f
  spectral implementation, 255–259
    nondimensionalization, 257
    results from, 257–259, 258f, 258t,
      259f
  tracking of interfaces, 250
  wavelet implementation and
    benchmarking, 259–268
    Cahn–Hilliard equation. *See*
      Cahn–Hilliard (1-D nonlinear
      diffusion) equation
    Daubechies wavelets, compactly
      supported, 260
    formulation, 261–264
    results from, 264–267, 265f, 265t,
      266f, 267t, 268f
Phase scalogram plots
  with secular variation plots and
    ridge function plots, 204, 207,
    209, 212, 213, 216
Phase transitions in earth,
    understanding of, 247–248, 249f

Piecewise constant approximation
(PCA), Haar wavelets and,
32–38
  box function in, 33, 33f
  Haar filter bank, 37
  Haar wavelet function, 35, 35f, 36
  on half interval, 34–35, 34f,
    35f, 37
  on unit interval, 32, 32f, 37
Poisson semigroup of wavelets, 170, 171,
  171f
Poisson wavelet family, for PF
  applications, 110–111
Polar region, GJ study in, 204–207,
  205t–206t, 207f
Polynomial matrix, defined, 45
Potential fields (PF), applications of WT
  to, 110–114
PPT (Pamatai), GJ study and, 206t, 209,
  211f, 213, 214f, 215f
P-QRS-T complex, ECG signals and, 143,
  144f, 145, 148
PR. *See* Perfect reconstruction (PR) filter
  banks
Product filter, PR filter banks design
  and, 43–44
  half-band filter, 44
Proteomic images, two-dimensional
  singularities in
  singularities detection (case study),
    148–153, 149f, 150f, 151f, 152f

Q

Quadrature mirror filters (QMFs), 103,
  104
  banks, 39
Quincunx filter banks, 149, 150f

R

Rainfall data using WT, analysis,
  234–238
  CWT, 235–236, 236f, 237f
  denoising data using DWT, 234
  Fourier analysis, 237–238, 238f
  time series plot, 234–235, 235f
Regularity condition, in wavelet
  functions, 6

Regularity measure, of smoothness,
  68–69
  holder regularity, 68
  Sobolev regularity, 68
Rescaled range *(R/S)* analysis, 157–159,
  158f
Ridge functions, estimation
  phase scalogram plots with secular
    variation plots and, 204, 209,
    212, 213
  using real wavelets, GJs events and,
    202–203, 203f
R/S (rescaled range) analysis, 157–159,
  158f

S

Savgol filter, in monsoon data filtering,
  223, 223f, 224t, 225
SBC (subband coding filter
  bank), 39
Scale factor, defined, 53
Scaling, property of fractals, 178–179,
  181
  multiscaling fractal sets, 182
Scaling coefficients, 54
Scaling exponent, defined, 11
Scaling function
  defined, 54
  wavelets and
    generating, 60–63, 63f, 64f, 65f
    smooth, convergence, 63–67, 65f,
      67f
    transition matrix, 64–67, 65f, 67f
Scaling region, defined, 180
Scaling spectral method, 13
Scalogram, in scale–space plane, 99
Second-generation wavelets (SGW),
  107–109
  advantages, 109
  lifting scheme, 108, 109
  spatial wavelet constructions for,
    109
  techniques for, 108
Secular variation plots
  ridge function plots and phase
    scalogram plots, 204, 207, 207f,
    208f, 209, 210f, 211f, 213
Seismic discontinuities, 248

Seismicity, multifractal analysis of,
        184–190, 185f, 186f, 187f, 188t,
        189f, 190f. *See also* Multifractals
        and multifractal analysis, in
        earthquake prediction
Seismotectonic map, multifractal
        analysis and, 181
Self-organized criticality (SOC)
    in distribution of seismicity, 189
Self-similar functions, 118–119, 133–138
    application, 135–138
        example, 137–138, 137f
    dyadic MRA axiom and, 121
    property of fractals, 178–179, 181
    scaling functions of MRA, examples,
        133–134, 133f
    WT of, 134–135, 134f, 135f
Set partitioning in hierarchical trees
        (SPIHT), 81, 118, 125, 126–127
SGW (second-generation wavelets),
        107–109
    advantages, 109
    lifting scheme, 108, 109
    spatial wavelet constructions for,
        109
    techniques for, 108
SHA (spherical harmonic analysis)
        technique, 197
Shift factor, defined, 53
Short-time Fourier transform (STFT),
        94, 95, 138
    nonstationary behavior in
        temperature anomaly and tree-
        ring data, 230
    windowed FT, calculation, 30
    WT *vs.*, 31
Short-time Hartley transform (STHT),
        94
Signal denoising, thresholding
        techniques for, 141–143, 142f,
        144f
Similarity, property of fractals, 178–179,
        181
Similarity dimension, defined, 179
Sine functions, in Fourier theory, 1–2
Singularities, 119, 120
    detection. *See* Singularities detection
    noise removal and, 138, 138f
    types of, 138, 138f

Singularities detection
    case studies, 143–153
        one-dimensional singularity in
            ECG signals, 143, 144f, 145–148,
            145f, 146f, 147f
        two-dimensional singularities
            in proteomic images, 148–153,
            149f, 150f, 151f, 152f
    wavelets in, 139–143
        denoising. *See* Denoising
        noisy signals, 140–141, 140f
        signal without noise, 139, 139f
Smoothing, wavelet
    PF data and, 112
Sobolev regularity measure, 68
SOC (self-organized criticality)
    in distribution of seismicity,
        189
Solid–liquid boundaries
    phase field modeling, evolution of.
        See Phase field modeling
Solid–solid boundaries
    phase field modeling, evolution of.
        See Phase field modeling
Source depth estimation
    WT of gravity data for, 170–172, 171f,
        172f
Southern hemisphere region, GJ study
        in, 205t–206t, 209, 211f
Spatial-orientation tree, defined,
        127–128, 127f
Spectral analysis method, 12–13
Spectral approach, in survey network
        design, 162–163. *See also* Survey
        network design
Speleothems, WT application to
        paleomonsoon data from,
        219–227
    Akalagavi cave, 220
    dating, 219
    defined, 219
    extracting low power, hidden cycles
        during wavelet analysis,
        222–225, 223f, 224f, 224t, 225f,
        226f, 227f
    formation, 220
    multitaper analysis, 220–222, 221f,
        222f
    stalagmite, 219

Spherical harmonic analysis (SHA)
　　technique, 197
Spherical harmonic models, in GJs, 198
SPIHT (set partitioning in hierarchical
　　trees), 81, 118, 125, 126–127
Stalagmite, speleothem, 219
STFT (short-time Fourier transform), 94,
　　95, 138
Subband coding, defined, 100
Subband coding filter bank (SBC), 39
Subband signals, 39
Successive approximation, defined, 32
SURE threshold, 143
Survey network design, 162–167
　　detectability limit of survey, 166–167,
　　　　167t
　　fractal dimension approach, 163, 164f
　　optimum grid, 164–165, 165f, 166f
　　spectral approach, 162–163
Synthesis equations, defined, 36
Synthesis filter bank, 39, 39f
Synthesis modulation matrix, defined,
　　42

T

Temperature anomaly, nonstationary
　　behavior in
　　wavelet perspective. *See*
　　　　Nonstationary behavior
TFP (time–frequency product), of
　　wavelets, 73, 74, 77–78, 79
　　minimizing, 80–81
Time center, defined, 76
Time–domain matrix–based design
　　methods, wavelet filter banks
　　design, 70–71
Time–frequency localization, 3
　　filter banks design, 79–80
Time–frequency product (TFP), of
　　wavelets, 73, 74, 77–78, 79
　　minimizing, 80–81
Time–frequency uncertainty principle,
　　76–79
　　continuous-time signal, 76–78
　　discrete-time signal, 78–79
　　　　direct extension from continuous-
　　　　time measure, 78–79
Time localization, defined, 2

Time series analysis
　　CWT uses in, 233
　　fractal analysis. *See* Fractal time
　　　　series characterization
　　fractals in, 11–16, 12f
　　　　density, of boreholes, 11–12, 12f,
　　　　　　13f
　　　　persistence of, 11
　　　　reflectivity sequence, of boreholes,
　　　　　　11–12, 12f, 13f
　　　　scaling spectral method, 13
　　　　spectral analysis method, 12–13
　　　　susceptibility distribution, of
　　　　　　boreholes, 11–12, 12f, 13f
Time series plot
　　extracting wavelet coefficients and
　　　　fitting on, 239, 239f, 240f
　　rainfall anomalies, 234–235, 235f
Time variance, defined, 76
Toeplitz matrix, defined, 64
Transition matrix, 64–67, 65f, 67f
Translational invariance, wavelet
　　properties, 7–8
Tree-ring data
　　nonstationary behavior in, wavelet
　　　　perspective. *See* Nonstationary
　　　　behavior
　　World Data Center for
　　　　Paleoclimatology, 240
Tree-ring width data using WT,
　　240–243
　　detrending, 240–241
　　wavelet analysis, 241–243, 241f, 242f,
　　　　243f
Tree structured filter bank, 51, 51f
Two-channel PR filter banks, 39–44
　　alias system function, 41
　　analysis modulation matrix, 42
　　maximally decimated, 40, 40f
　　nonideal analysis filters of,
　　　　40, 40f
　　synthesis modulation matrix, 42
Two-dimensional singularities in
　　proteomic images
　　singularities detection (case
　　　　study), 148–153, 149f, 150f,
　　　　151f, 152f
Two-dimensional wavelet transform
　　(2D-WT), 104–107

U

Uncertainty principle
  Heisenberg's, 95–96
  time–frequency, 76–79
    continuous-time signal, 76–78
    discrete-time signal, direct
      extension from continuous-
      time measure, 78–79
    Heisenberg, 76

V

Valid wavelet filter banks, defined, 59
Vanishing moments, of wavelets, 7

W

"Watershed," image segmentation
  method, 150
Wavelet analysis, GJs and, 201–204
  complex wavelet analysis of, 204
  CWT, 201–202
  DWT, 201–202
  estimation of ridge functions using
    real wavelets, 202–203, 203f,
    212, 213, 216
Wavelet coefficients
  extracting, fitting on time series plot,
    239, 239f, 240f
Wavelet-collocation methods
  for implementing phase field model,
    248, 259–268. *See also* Phase
    field modeling
    Daubechies wavelets, compactly
      supported, 260
    formulation, 261–264
    results from, 264–267, 265f, 265t,
      266f, 267t, 268f
"Wavelet families," defined, 6
Wavelet filter banks design methods,
    69–71
  factorization of LHBP, 69–70
  lattice structures, 70
  lifting scheme, 70
  parametrization of filter coefficients,
    70
  time–domain matrix–based design,
    70–71

Wavelet frames, defined, 99
Wavelet packet transform (WPT),
    102–103
Wavelet ridges, of WT, 171
Wavelets
  advantages, 96
  analysis, tree-ring width data,
    241–243, 241f, 242f, 243f
  analyzing (mother) wavelet, 6
  in applied geophysics, 155–173
    matrix multiplication, for
      gradient-based inversion,
      167–170, 169f
    overview, 155–156
    WT of gravity data for source
      depth estimation, 170–172, 171f,
      172f
  basic theory and mathematical
    concepts, 5–11
  and complex wavelets in
    geomagnetism, 199. *See also*
    Geomagnetic jerks (GJs)
  construction, 30–90
    CWT *vs.* DWT, 52–53
    design methodology, 80–90, 82t,
      83t, 84f, 85f, 85t, 86t, 87t, 88f,
      89f, 90f
    filter banks. *See* Filter banks
    Haar wavelets, PCA and. *See* Haar
      wavelets, construction
    orthogonal wavelets, design. *See*
      Orthogonal wavelets design
    overview, 30–32
    parametrization technique, 73–76
    time and frequency conflict,
      uncertainty principle, 76–79
    time–frequency localization,
      79–80
  CWT. *See* Continuous wavelet
    transform (CWT)
  defined, 4, 5
  DWT. *See* Discrete wavelet transform
    (DWT)
  filter banks and, connection, 54–69
    condition for convergence in $L_2(R)$,
      63–67, 65f, 67f
    FWT, 57–59, 59f
    generating scaling functions and,
      60–63, 63f, 64f, 65f

MRA. *See* Multiresolution
    analysis (MRA)
  PR filter banks and, 59–60
  regularity measure, 68–69
  transition matrix, 64–67, 65f, 67f
  fractals and, 18–20
    wavelet-based fractal analysis, 20
  in geophysics, 16–17
  implementations of phase field
      methods, 248, 259–268. *See also*
      Phase field modeling
    Daubechies wavelets, compactly
      supported, 260
    formulation, 261–264
    results from, 264–267, 265f, 265t,
      266f, 267t, 268f
  overview, 1–4
  Poisson semigroup of, 170, 171, 171f
  properties, 7–8
    compact support, 7
    translational invariance, 7–8
    vanishing moments, 7
  reasons for, 96–97
  SGW, 107–109
  in singularities detection, 139–143
    denoising. *See* Denoising
    noisy signals, 140–141, 140f
    signal without noise, 139, 139f
  wavelet functions, 5–6
Wavelet transform modulus maxima
    (WTMM), 139
Wavelet transforms (WT), 52–53
  application to paleomonsoon
      data from speleothems. *See*
      Paleomonsoon data from
      speleothems
  in applied geophysics
    gravity data for source depth
      estimation, 170–172, 171f, 172f
  genesis, types and applications,
      93–114
    CWT. *See* Continuous wavelet
      transform (CWT)

DWT. *See* Discrete wavelet
    transform (DWT)
  FWT, 103–104
  Heisenberg's uncertainty
      principle, 95–96
  history, 97
  overview, 93–94
  to potential fields, applications,
      110–114
  SGW, 107–109
  two-dimensional, 104–107
  wavelets, reasons, 96–97
  WPT, 102–103
  nonstationary behavior in
      temperature anomaly and tree-
      ring data, 230–231
  analysis of rainfall data using. *See*
      Rainfall data using WT
  extracting wavelet coefficients
      and fitting on time series plot,
      239, 239f, 240f
  tree-ring width data using. *See*
      Tree-ring width data using WT
  nonstationary signals, analysis,
      30–31
  of self-similar functions, 134–135,
      134f, 135f
  STFT *vs.*, 31
Weierstrass, Karl, 4
Wen Liang Hwang, 139
Wigner–Ville distribution (WVD), 30
Windowed FT (WFT), 3
Wingst (WNG) observatory, GJ study
    and, 206t, 208f, 209, 212, 213,
    214f
World data center (WDC-C2) for
    geomagnetism, 199, 200
World Data Center for Paleoclimatology,
    240
WPT (wavelet packet transform),
    102–103
WTMM (wavelet transform modulus
    maxima), 139

T - #0391 - 071024 - C308 - 234/156/14 - PB - 9780367379193 - Gloss Lamination